An excitin und-up
of the world's most
exhilarating cars.

LRX 4

CLASSIC SPORTS CARS

GRAHAM ROBSON

PSL

Patrick Stephens, Wellingborough

First published in 1986

British Library Cataloguing in Publication Data

Robson, Graham
Classic sports cars.
1. Sports cars—History
I. Title
629.2'222 TL236

ISBN 0-85059-839-7

Colour photographs (including cover) from Classic and Sportscar *and* Autocar, *with thanks to Haymarket Publishing Limited.*

Patrick Stephens Limited is part of Thorsons Publishing Group

Printed in Great Britain by Woolnough Bookbinding, Irthlingborough, Northants

2 4 6 8 10 9 7 5 3

CONTENTS

INTRODUCTION

There is nothing new about the sports car. Many years ago, even before the beginning of the twentieth century, there were cars which certainly had a sporting character, and this pedigree developed strongly in the years which followed. Then, in the 1930s, the Vintage Sports Car Club was founded, the phrase 'vintage car' was invented, and soon everyone knew what a 'vintage sports car' looked like. In due course, the 'thoroughbred sports car' made itself obvious, and the world of motoring was happy about this for a number of years.

In the USA, too, the 'Milestone' car was defined, and the Classic Car Club of America also decided to recognize a series of fine machines built between 1925 and 1948. Are you still with me, because complications are about to set in!

This left a huge hole, into which hundreds of thousands of fine post-war cars fell into oblivion. They were never going to be 'vintage' because they were far too young, nor 'thoroughbred' unless the VSCC said they were, and they could never become 'American Classic' if they were built after 1948, and unless the patrician CCCA invited them to join in.

In the 1960s and early 1970s, a new movement developed. Enthusiasts all over the world, particularly in Europe, who loved post-war cars of character, began trying to categorize this more modern machinery, which fitted neither the 'vintage' nor simply the 'thoroughbred' group. The definition eventually chosen, for cars as different as the Ferrari 250GT and the Triumph TR3A, the Morgan Plus 8 and the Porsche 911, was 'classic', which was an entirely different grouping from that of the CCCA!

Thus, the European 'classic car' movement was born, and it has been expanding ever since. This book is an attempt to list a selection of the most important 'classic sports cars' built since 1945.

What is a 'classic sports car'?

I'd better make a confession, straight away. The pleasure of producing a book about classic sports cars has not been completely trouble-free. Some of the cars included are famous enough, but some equally well-known models have had to be omitted. There must have been a reason. There was. In years gone by, they say, every budding motoring writer had to produce a convincing definition of the phrase 'sports car', before he won the respect of his peers.

That was — and is — bad enough, but in the 1980s there is now an equally difficult problem. How does one define the phrase 'classic car'? To compound it all, therefore, what on earth is a 'classic sports car'?

You see my problem? I have been driving, enjoying, and writing about fast, beautiful, and exciting cars for a good many years, and I have learned that no two pundits will agree on everything. With that in mind, I hope that the reader will accept, and admire, the choice of cars included in this survey, for there was a practical limit to the numbers, and the detail about them, which I could use.

What is a 'sports car', and what is a 'classic car'? In both cases, I believe, the individual motoring enthusiast will be sure of his own definition, and this is quite likely to differ from mine. But I must still offer my own...

I will start by stating what a 'sports car' is not. It is not necessarily a two-seater, and it is not necessarily an open car. It does not need to have an exotic technical specification, and it certainly need not have tremendous performance or a high price tag.

A 'sports car', on the other hand, must be for fun. It must make the driver feel a hell of a guy. It must make the driver feel that his or her skills have suddenly improved dramatically, that there is no reason why race or rallying victories should not follow at once, and that there has never been as much motoring enjoyment before. It must, in other words, make the driver *feel* sporting because of its character.

Character! That's the word I have been looking for — or one of them, at least. Anyone can design a fast car, a two-seater car, an open car, or something called a 'sports car', but if it does not have character it will never be accepted as such. You wouldn't call a 'Frog-eye' Sprite a fast car, but you would most certainly call it a sports car. On the other hand, you might admire a Rolls-Royce Corniche for many things, not least its performance, and its looks, but its character is definitely non-sporting.

That makes it easier. If you can get agreement that a particular motor car meets certain performance, handling and styling standards, you are part of the way there. If the same car has that indefinable touch of character built in, then the problem is solved. You have found a 'sports car'.

Examples? Certainly. By my standards, an MG MGB GT is a 'sports car', even though it was never very fast, and in some years did not have very distinguished handling. A Ford Capri 3 litre GT, on the other hand, somehow misses the 'sports car' title, even though it was faster, more roomy, and more practical than the MGB GT. Sometimes, though, relatively minor differences can change a car completely — consider the 2.8 Injection Capri, and compare it with the 3-litre. See what I mean?

Incidentally, a 'proper' sports car certainly does not need to have specially-designed running gear, though naturally it helps. Many a fine car uses mechanical items from other machines, some of it surprisingly mundane. The 1950s BMC sports cars — MGAs and Austin-Healeys — were perfect examples of models making the best out of engines and even suspension sub-assemblies originally designed for use in an ordinary saloon car. Corvettes use the same engines and transmissions as any other Chevrolet, and even the most luscious Alfa Romeos sometimes used engines and gearboxes also found in saloons or even light commercial vehicles.

By my standards, a sports car can certainly have more than two seats — shall we say generous 2+2-seater accommodation? — and I sometimes have difficulty in eliminating full four-seaters. The Porsche 928, for instance, is nearly a saloon car according to the interior measurements, but it is surely still a sports car.

A sports car certainly does not need to be an open car, or even a fully-trimmed cabriolet. Something like the Chevrolet Corvette of the 1970s, the Porsche 911, or the Lamborghini Miura *must* qualify as a sports car, even though all are normally built with a solid roof panel.

All in all, therefore, any knowledgeable lover of motor cars will recognize a proper sports car when he sees one, and he will immediately reject the bogus, or even the 'near miss'. I hope that everyone studying the list of cars I have included in this book will agree with me.

'Classic' — does it mean more?

So far, so good, but now we have to talk about a 'classic'. I think I should start, therefore, by looking at a dictionary definition, which states that 'classic' is: 'Of the first class, of avowed excellence', while the 'classic style' is defined as: 'simple, harmonious, proportioned and finished'.

I think this is going to be easier, relatively, than sorting out the sports car definition. I think we can also say that almost every respected sports car automatically qualifies for 'classic' status. On the other hand, there are dozens of classic cars ('of the first class, of avowed excellence', remember...) which are not sports cars. A Rolls-Royce Phantom VI, for sure, is classic, but I'm sure it would not pretend to be a sports car. Neither would a Wankel-engined NSU Ro80, a Jeep, or a VW Beetle, though I insist that they are all 'classic', of their type.

To qualify as a 'classic', a sports car must have all the qualities already listed, but it must have more. It must have a definite personality, and that indefinable quality which I can only define as 'style'. Many cars make motoring safe, rapid, comfortable and even enjoyable, but some turn a journey into a real pleasure, even a special occasion.

What is interesting is that some 'classic' sports cars take on that title years after they were born. In the motoring business, they have often been what we call 'sleepers' for a time — it is only after a lot of experience that they become so endearing, and so attractive, that they begin to look 'classic'. A fine car like a Lancia Stratos, a Giulietta Sprint, or a Jaguar XK120 is recognized instantly for what it is, but a car like the Fiat 124 Spider, or the MG TF looks better as hindsight takes over.

Many classic sports cars, I have to agree, were costly and exotic when new, and have taken on real rarity value since then, but there are others which sold in tens or even hundreds of thousands at much more modest levels of performance. No matter. Whatever the price tag when they were new, these 'classics' were never ordinary, never totally predictable, and never ever boring to own or drive.

One interesting facet of the breed is where the vast majority of them have been made, and where most were sold. Most classic sports cars have been designed, if not always manufactured, in Europe, while most of them were sold, when new, in North America. The 'home' of the modern-day classic was in Europe because it was only in countries like Italy, West Germany, France and Britain that the love of such cars truly existed, and where there were so many sporty-minded managers ready to indulge their designers and stylists. Most sales were in North America, quite simply, because the USA was, and is, the most prosperous nation on earth, with the largest motor car market of all.

Amazingly enough, the United States motor industry has rarely attempted to build such cars, with the Chevrolet Corvette standing alone until the 1980s, when it was joined by the Pontiac Fiero. Both were produced by General Motors companies: neither Ford nor Chrysler have ever joined in. The only other large car-producing nation, Japan, entered very late, with the Datsun 240Z, but is now catching up fast. In future editions of this book, it is highly likely that there will be more candidates from Japan.

Twenty years ago, only the CCCA had defined 'classic', and then only by their own exclusive (and 'expensive') standards. In Europe, it was the onset of what I call 'Detroitisation' (the production of whole new ranges of cars, with different engines, different names, and even different countries of origin) which drove much of the character out of our motor cars. Motoring enthusiasts were therefore forced to look around much more carefully for what new sports cars were left, or to buy, love, and restore ageing cars from earlier decades.

Accordingly, we can blame Big Business, legislation and 'modern times' for the flourishing development of the classic car cult. From the early 1970s, when the first British 'classic' magazine, *Thoroughbred & Classic Cars*, was founded, interest has grown and grown, and a remarkable number of long-neglected cars have had their reputations revived. In particular, dozens of new 'one make' car clubs have been founded, and some have grown very large indeed. The British MG Owners' Club was first to reach the 50,000 membership level: it will not be the last. Nowadays there are races, rallies, concours events, trials and mere social get-togethers, all devoted to the continuing phenomenon, the classic car.

All of which has left me with a real dilemma — what to include, and what to omit. I admit freely that there may be cars which should have been surveyed in the following pages, and I also admit that I would have liked a lot more space to describe the merits of those which *have* found a place. For that reason I have grouped cars together as ranges, or families — for example, all the E-Types, six-cylinder and V12, have been put together, as have all the Porsche 911s, and all the separate-chassis Triumph TRs.

Do you admire a particular classic sports car? Would you like to own one? Read on...

Graham Robson
Bridport, Dorset, February 1986

Note In the model descriptions which follow, I have chosen to display a specification summary for the most significant version only. Where possible, I have added a comparison with other models in the range.

AC ACE AND ACECA

Produced At Thames Ditton, UK, 1953 to 1964. Production: 693 Ace, 350 Aceca (and 83 2+2 Greyhound Coupés). **Data for Ace-Bristol** Tubular chassis, with light-alloy body shell. Front-mounted Bristol ohv straight-six engine driving rear wheels. 1,971cc, 105 to 130 bhp at 4,750/6,000 rpm depending on chosen tune. Four-speed gearbox. All-independent suspension (transverse leaf, wishbones, front and rear); worm and sector steering; front disc/rear drum brakes (all drums, early cars); 5.50-16in tyres. Wheelbase 7ft 6in; overall length 12ft 8.5in; overall width 4ft 11in. Unladen weight 1,720 lb. **Derivatives** Ace-AC engine had 1,991cc, 85 to 102bhp at 4,400/5,000 rpm. Aceca was a fastback coupé version of the Ace, with all engine options, and 13ft 4in length, 1,990 lb weight. (Greyhound looked similar, but had longer wheelbase, different style, different chassis and suspensions.) **Performance potential** Ace-Bristol: Top speed (125 bhp engine) 117 mph; 0-60mph 9.1 sec; standing ¼-mile 16.5sec; typical fuel consumption 23 mpg (Imperial). Aceca-AC: Top speed (90 bhp engine) 102 mph; 0-60mph 13.4sec; standing ¼-mile 19.1sec; typical fuel consumption 22 mpg (Imperial).

Not only was the Ace family a fine design in its own right, but it was also famous as the predecessor of the V8-engined Cobra which evolved from it, and for its use of the 'Barchetta' body style of the period. The car, in fact, was developed by AC from the basic design of a British Tojeiro racing sports car of the period, and had no chassis links with the AC 2 litre touring cars which it replaced.

The original car was the open two-seater, with the light-alloy six-cylinder AC engine, but this unit had a truly ancient heritage, and was close to its development limits. In 1956, therefore, AC offered the BMW

The timeless style of the AC Ace, derived from the Tojeiro shape of the early 1950s.

The Aceca was really the Ace with a permanent coupé roof, but the same chassis and performance.

328-derived Bristol engine as an alternative, for this was a more modern, and tunable, unit.

The Ace was joined by the fastback Aceca coupé in 1955, both cars being based on the same, simple, tubular frame, with equally simple but effective all-independent suspension. Bodies were in light alloy, with some glass-fibre inner panels, and all were strictly two-seaters.

The Greyhound of 1959 looked similar to the Aceca, but had an entirely different chassis and rear suspension, with longer wheelbase and wider tracks. It was only ever available with the Bristol engine.

The last AC-engined car was built in 1961, at almost the same time that Bristol stopped making their own engines, so AC had to find an alternative. This led to the launch of the Ace and Aceca 2.6 derivatives, where the engine was the straight six from the British Ford Zephyr, supplied in several states of tune, and they were not a success.

Even so, all these cars were possessed of splendid good looks, and had excellent roadholding by the standards of the day, though we might now laugh at the narrow tyres and the strange camber angles taken up by the wheels under hard cornering. They were — and are — rare, not only because they were always expensive to buy and run, but because AC were never equipped to build cars in large numbers.

Production of 2.6s was running down rapidly before the Cobra arrived — and the American-engined car soon dominated the scene at Thames Ditton.

AC COBRA

Produced At Thames Ditton, UK, and completion at Los Angeles, USA, 1962 to 1969. Production: 75 260 cu in cars, 580 leaf-spring 289s, 356 427s, and 27 AC 289s (427 chassis/289 engine). **Data for Cobra 289** Tubular chassis, with light-alloy body shell. Front-mounted Ford-USA ohv V8 engine driving rear wheels. 4,727cc, 195 bhp (several options) at 4,400 rpm. Four-speed gearbox. All-independent suspension (transverse leaf, wishbones, anti-roll bars, front and rear); rack-and-pinion steering; four-wheel disc brakes; 7.35-15in tyres. Wheelbase 7ft 6in; overall length 13ft 2in; overall width 5ft 3in. Unladen weight 2,315 lb. **Derivatives** Original Cobra 260 had 4,260cc, 164 bhp at 4,400 rpm, plus worm-and-sector steering. Cobra 427 had 6,998cc or 7,013cc, 345 bhp (several options) at 4,600 rpm, 2,530 lb weight. Coil spring, wishbone suspension. Width 5ft 8in. AC 289 had 427 chassis, with 289 engine. **Performance potential** Cobra 289 (271 bhp, gross, engine): Top speed 138 mph; 0-60mph 5.5sec; standing ¼-mile 13.9sec; typical fuel consumption 17 mpg (Imperial). Cobra 427: Top speed 143 mph; 0-60mph 4.8sec; standing ¼-mile 12.9 sec; typical fuel consumption 12mpg (Imperial).

Carroll Shelby, who inspired the design of the Cobra, but most certainly did not design it, had his bright idea in 1961, and the first cars were ready for sale during 1962. Shelby's brainwave, which made him, and AC, a lot of money, and produced a car of great charisma if controversial technical merit, was to see a large-capacity Ford USA V8 engine shoehorned into a suitably strengthened AC Ace chassis with open two-seater body shell. It was only feasible because the chosen engine was light and compact, and it was enormously successful because the

The early Cobra looked very similar to the Ace, except for the wider wheels and tyres.

combination of massive Detroit-style torque and light vehicle weight produced a sports car of colossal performance, and great character.

The engines and transmissions, in fact, were fitted at Shelby's premises in California, the cars being shipped out to him, part-completed, from the UK, and the vast majority of sales were in the United States. At first the original Cobras had 4.2 litres and worm-and-sector steering, but a change was made to 4.7 litres after 75 cars had been built, and rack-and pinion steering was also adopted in 1963.

Those cars became known, retrospectively, as Cobra Mk IIs, or 289s, in 1965, when a radically redesigned version was introduced, complete with coil spring independent suspension, and the even more massive 7.0 litre Ford V8 engine — the '427' referring to the engine's capacity in cubic inches. This was the car said to have had 'computer-designed' suspension geometry — though considerable changes had to be made by AC's designers to make it feasible in a road car!

No-one, not even the greatest Cobra enthusiast, would ever call the car refined, or even properly-developed, for its acceleration and performance far outweighed its roadholding and general habitability. Latter-day cynics, indeed, have come to call the Cobra an 'obscene' car for the way it behaves — and there is no doubt that it appeals primarily to extroverts.

The short-lived European version, was known as the AC 289.

Once the Cobra had its 7-litre engine, its flared arches, and its ultra-fat tyres, it was an unmistakable monster.

ALFA ROMEO GIULIETTA

Produced At Milan, Italy, 1954 to 1965. Production: 27,142 Sprint and Sprint Veloce, 17,096 Spider and Spider Veloce, 1,366 Sprint Speciale, 200 SZ. **Data for Sprint** Unit-construction pressed-steel body/chassis unit. Front-mounted 2 ohc four-cylinder engine, driving rear wheels. 1,290cc, 80 bhp at 6,000 rpm. Four-speed gearbox. Coil spring independent front, coil spring live axle rear suspension; worm and sector steering; four-wheel drum brakes; 155-15in tyres. Wheelbase 7ft 9.7in; overall length 12ft 7.5in; width 5ft 0.5in. Unladen weight 1,935lb. **Derivatives** Spider had open two-seater style, on 7ft 4.6in wheelbase. 'Veloce' engine option had 90 bhp at 6,000 rpm. Sprint Speciale and Sprint Zagato were Bertone and Zagato styled coupés on Spider floorpan, with 100 bhp at 6,500 rpm, and five-speed gearbox. Weights 1,895lb and 1,700 lb, respectively. **Performance potential** Sprint: Top speed 101 mph; 0-60mph 13.0sec; standing ¼-mile 18.9sec; typical fuel consumption 28 mpg (Imperial). SZ, with Zagato body: Top speed 122 mph; 0-60mph 11.2sec; standing ¼-mile 17.8sec; typical fuel consumption 27 mpg (Imperial).

It was the Giulietta, and in particular its famous twin-cam engine, which inspired several generations of more modern Alfa Romeos, so it is fitting that the original car is still so well-loved. Until the second World War, Alfa Romeo built only limited numbers of costly and exotic motor cars, but the scene in Milan was transformed in post-war years. The state-controlled company set out on a strategy of massive expansion, first with the 1900 saloons and then — from 1954 — with a range of cars we now know as the Giuliettas.

It was an ambitious programme, for every component in the design was brand new, and there were to be saloons, coupés and spiders in the range, while the engine was even planned to be used in light commercial vehicles too. The basic model from which all others were developed was a lively, but rather high and narrow, four-door saloon, and the power unit was a 1.3 litre twin overhead camshaft engine.

The Giulietta saloon, in fact, was not revealed until 1955, by which time the legendary Bertone-styled sprint coupé, a four-seater which used the same floorplan and the same running gear, had been on sale for a year. Bertone also built the shells, and sent them to Milan for completion — this being an ideal way to shake down an entirely new 'chassis' and wring out the teething troubles before the mass-market saloon arrived.

The pretty little open two-seater Spider, introduced in 1955, was styled, and the body was produced, by Pininfarina on a 5.1 inch shorter version of the floorpan, but otherwise used the same chassis.

Coupé and Spider models originally had 80 bhp, but from 1956 they were also available as 'Veloce' versions, with 90 bhp. The most exciting derivatives of all, however, were the Bertone-styled Sprint Speciale, and the Zagato-styled SZ coupés, which used the shorter-wheelbase floorpan, were much lighter, and had 100 bhp, plus five-speed transmissions. The SZ, in particular, was a successful competition car.

Above *Bertone's exquisitely-detailed Giulietta Sprint coupé made twin-cam motoring available to the middle-classes.*

Below *The Giulietta Spider was shaped by Pininfarina, and had a shorter wheelbase than the Sprint.*

ALFA ROMEO GIULIA

Produced At Milan, Italy, 1962 to date. Production: 231,328 Sprint coupés of all types, 63,000 Spider/Convertibles (to end '83), 1,400 SS, 112 TZ, 962 GTAs of all types, 1,510 Zagato-bodied types. **Data for Sprint 2000GTV** Unit-construction pressed-steel body/chassis unit. Front-mounted 2 ohv four-cylinder engine, driving rear wheels. 1962cc, 131 bhp at 5,500 rpm. Five-speed gearbox. Coil spring independent front, coil spring live axle rear suspension; recirculating ball steering; four-wheel disc brakes; 165-14in tyres. Wheelbase 7ft 10.5in; overall length 13ft 5in; width 5ft 2in. Unladen weight 2,290 lb. **Derivatives** Spider had open two-seater style, on 7ft 4.6in wheelbase. Convertible was 4-seater DHC conversion of Sprint, SS was Giulietta SS body style, TZ was tubular-framed racing-sports version, GTA was light-alloy shelled Sprint, Zagato models were specially styled on Spider wheelbase. Engines, over the years, varied from 89 bhp 1,290cc, through 1,570cc and 1,779cc, to 1,962cc. **Performance potential** 2000 GTV: Top speed 120 mph; 0-60mph 9.2sec; standing ¼-mile 16.4sec; typical fuel consumption 23 mpg (Imperial). 1300 (Sprint) Junior: Top speed 102 mph; 0-60mph 13.2sec; standing ¼-mile 19.1sec; typical fuel consumption 27 mpg (Imperial).

In 1963, Alfa Romeo achieved a very difficult target — they replaced the extremely successful Giulietta family by an even better and more attractive range, called the Giulia. In essence, Giulias were bigger, better, faster and more modern-looking Giuliettas, and drew on all the experience — and reputation — built up with the original cars.

As before, it was the new Giulia saloon which was intended to sell in the largest numbers, while the same styling and production agreement was reached with Bertone and Pininfarina for the coupés and open spiders. The basic 'chassis' of the latest cars was much as before,

Bertone's Giulia Sprint GT was even more elegant than the Giulietta which it replaced.

In its original long-tail form, the Giulia Spider was called 'Duetto' —Pininfarina styled it.

though the engines were enlarged to 1.6 litres (and would get progressively bigger in the years to come). Five-speed gearboxes were standard.

To thoroughly confuse everyone, incidentally, there were 'interim' Giulias, where the new running gear was grafted into the Giulietta-style coupés and spiders, though this arrangement only lasted from 1962 to 1965...

The first true Giulia coupé was the Sprint GT of 1963, while the controversially-styled Pininfarina Duetto spider arrived in 1966. There was also a short-lived convertible version of the Sprint GT called the GTC (which still had four seats, and was a conversion by Carrozzeria Touring), GT Veloce stages of tune with 109 bhp from 1966, and a 115 bhp lightweight GTA (A=Aluminium) version of the GT from 1965. Life was never dull for production planners at Alfa Romeo!

In the late 1960s and early 1970s a whole series of changes — mechanical and styling — were made, including the offering of different engines (some as large as 2 litres, some as small as 1.3 litres), a chopped-off rear for the Spider, which accordingly lost the 'Duetto' title, and a smart new Zagato coupé style on the shortest version of the floorpan.

The last of the coupés was produced in the mid-1970s, but Spider assembly continued through to the mid 1980s, to keep the USA market satisfied.

ALPINE-RENAULT

A108/A110 — **Produced** At Dieppe, France, 1957-77. Production: of A108s not known, 7,160 A110s **Data for A110 1300** Tubular backbone chassis frame, with glass-fibre body shell. Rear-mounted Renault ohv four-cylinder engine and transmission driving rear wheels. 1296cc, 110 bhp at 6,900 rpm. Five-speed gearbox. Coil spring independent front, coil spring and swing axle independent rear suspension; rack and pinion steering; four-wheel disc brakes; 135-380 tyres. Wheelbase 6ft 10.7in; overall length 12ft 7.6in; width 5ft 1in. Unladen weight 1,200 lb. **Derivatives** A108s had smaller Dauphine-type engines, 845cc to 997cc. Coupés, cabriolets and 2+2 seater versions all available — but 'Berlinette' Coupé much the most numerous. A110 engines from 956 to 1,647cc, of two basic types, up to 127 bhp. From 1973, rear suspension was by wishbones. **Performance potential** A110 1300: Top speed 123 mph; 0-60mph 9.1 sec standing ¼-mile 16.6sec; typical fuel consumption 30 mpg (Imperial).

The Alpine-Renault *marque* was invented by Jean Redelé of Dieppe, and the cars have always been built in that French town. Each and every one has used rear-mounted Renault engines and transmissions, glass-fibre body styles, and all-independent suspension.

The first cars were Type A106s, using Renault 4CV undersides and running gear, but the famous A108/A110 models of 1957-77 used a tubular backbone frame, and had rakish styling. From the mid-1960s, with larger five-bearing engines, they were successful race and rally cars, mostly being built in fastback coupé guise.

From 1971, too, there was the smart A310 variety, much more of a road car than before, and this was replaced by a new Renault-designed car in 1985.

All the best Alpine-Renaults were Berlinettas — this being the A110 version of the late 1960s.

ASTON MARTIN DB2 FAMILY

Produced At Feltham, UK, and latterly at Newport Pagnell, UK, 1950 to 1959. Production: 411 DB2s, 764 DB2/4s, 551 DB Mk IIIs — all body types. **Data for DB2/4 2.9-litre** Multi-tubular chassis frame, with steel and light-alloy body shell. Front-mounted 2 ohc six-cylinder engine, driving rear wheels. 2922cc, 140 bhp at 5000 rpm. Four-speed gearbox. Coil spring independent front, coil spring live axle rear suspension; worm and roller steering; four-wheel drum brakes; 6.00-16in tyres. Wheelbase 8ft 3in; length 14ft 1.5in; width 5ft 5in. Unladen weight 2,770 lb. **Derivatives** DB2 was original car, with 2,580cc/105 bhp. DB Mk III followed DB2/4 with 162 or 178 bhp engine, front wheel disc brakes. Convertible versions were available, all types, and hardtop DB2/4 Mk IIs and DB Mk IIIs. **Performance potential** DB2/4 2.9-litre: Top speed 119 mph: 0-60mph 11.1sec; standing ¼-mile 17.9 sec;' typical fuel consumption 21 mpg (Imperial).

Until the end of World War Two, Aston Martin and Lagonda were both independent companies, each with famous reputations and building cars in very limited quantities. Then, in the severe conditions of post-war austerity, both fell on hard times and were taken over by tractor industrialist David Brown who — as he later admitted — was 'looking for a bit of fun'. It was his inspired decision to mate a new Lagonda engine with a new Aston Martin chassis which gave birth to the very successful DB (or 'David Brown') series of cars.

The multi-tubular Aston Martin chassis, complete with trailing arm independent front suspension and a properly-located back axle, had been designed by Claude Hill to use a new four-cylinder engine, while the twin-overhead-camshaft six-cylinder Lagonda engine had been designed under the leadership of W.O. Bentley for use in a new and compact luxury saloon. The saloon eventually *did* go on sale, but the alternative engine was abandoned after a very short career in the DB1.

The DB2 was first raced as a prototype in 1949, but the first production cars were not delivered until 1950, at which time the only available body style was a fastback two-seater coupé, with no external boot access and only two seats. The body was still somewhat fragile, but beautifully styled if rather simply furnished. Within months an alternative drop-head coupé became available. Final assembly of both cars was at Aston Martin's 'traditional' home of Feltham, in Middlesex.

The DB2 was a great car with splendid roadholding and a 110 mph top speed from its 2.6 litre/105 bhp engine, but the DB2/4 which followed in autumn 1953 was even better. Not only did it have a reshaped body on the same 8ft 3in wheelbase, with an opening rear hatchback, but it also had extra '+2' seating, and shells were supplied by Mulliners of Birmingham (they were built alongside the original Triumph TR2 bodies, which were also made there).

A year later the 'W.O.' engine was enlarged to 2.9 litres, producing 125 bhp, which raised the top speed to nearly 120 mph and there was also a drop-head coupé version of the body. A year after that, with the Tickford coachbuilding business now also owned by David Brown, the body contract was taken away from Mulliners and the Mk II version of the DB2/4 arrived, looking visually very similar.

There was still more to come from the design, which progressed to become the DB Mk III (not to be confused with the DB3, a sports racing Aston Martin). In that guise the car had a more delicately styled nose, yet more power (some engines, optionally, gave up to 178 bhp), and the choice of front-wheel disc brakes, which became standard after the first 100 examples had been built. The final series of Mk IIIs were completely assembled at Newport Pagnell, where Aston Martin production was progressively concentrated in the late 1950s. It later gave way to the much larger, and less sporting, DB4 model.

Although this DB series never had the same brute performance and 'sex-appeal' as the contemporary Jaguar XK models, it always had better roadholding, arguably nicer styling, and a more discerning clientele. Hand-building, however, costs money, which explains why the cars were always expensive when new, and why they are now so costly to preserve for posterity.

Below *The DB2 was a two-seater, and the whole of the bonnet and front wings could be lifted up together.*

Top right *The DB2/4 was available as a coupé, or as a drop-head model, both of which are highly-prized cars today.*

Centre right *Though not wind-tunnel-tested, the DB2 had a very smoothly shaped tail.*

Right *The final derivative was the DB Mk III, with distinctive nose style, disc brakes and more power than ever.*

WHM 555

WLY 98

AUSTIN-HEALEY 100 FAMILY

Austin-Healey 100 (four-cylinder) family — Produced At Longbridge, Birmingham, UK, 1953 to 1956 (100S and 100M completed at Warwick, UK). Production: 10,688 BN1s, 3,924 BN2s, 50 100Ss. (100M figures hidden in BN1/BN2 totals.) **Data for 100 BN1** Box-section chassis frame, with steel body welded to it on assembly. Front-mounted ohv four-cylinder engine, driving rear wheels. 2,660cc, 90 bhp at 4,000 rpm. Three-speed gearbox, with overdrive. Coil spring independent front, leaf-spring live axle rear suspension; cam and peg steering; four-wheel drum brakes; 5.90-15in tyres. Wheelbase 7ft 6in; overall length 12ft 7in; width 5ft 0.5in Unladen weight 2,150 lb. **Derivatives** 100S was much-lightened version of BN1, with 132 bhp at 4,700 rpm, and four-wheel disc brakes. BN2 was like BN1, but with four speeds and overdrive. 100M was modified BN2, with 110 bhp at 4,500rpm. **Performance potential** BN1: Top speed 103 mph; 0-60mph 10.3 sec; standing ¼-mile 17.5 sec; typical fuel consumption 25 mpg (Imperial).

Austin-Healey (six-cylinder) family — Produced At Longbridge, UK and (from late 1957) at Abingdon, UK, 1956 to 1968. Production: 14,436 100 Six, 13,650 3000 Mk I, 5,450 Mk II, 6,113 Mk II Convertible, 34,025 Mk III. **Data for 3000 Mk III** Box-section chassis frame, with steel body welded to it on assembly. Front-mounted ohv six-cylinder engine, driving rear wheels. 2,912cc, 148 bhp at 5,250 rpm. Four-speed box with optional overdrive. Coil spring independent front, leaf-spring live-axle rear suspension; cam-and-peg steering; front discs, rear drum brakes; 5.90-15in tyres. Wheelbase 7ft 8in; length 13ft 1.5in; width 5ft 0.5in. Unladen weight 2,548 lb. **Derivatives** 100 Six was first six-cylinder car, with 2,639cc/102 bhp (later 117 bhp), and drum brakes; 3000 Mk 1 had 2,912cc/124 bhp, front discs; Mk II had 132 bhp; Mk II Convertible was first to have fully-folding hood. **Performance potential** 3000 Mk III: Top speed 121 mph; 0-60mph 9.8 sec; standing ¼-mile 17.2 sec; typical fuel consumption 24 mpg (Imperial).

The Austin-Healey 100 of 1953/54 had a four-cylinder engine. The hardtop was an option.

By 1959 the six-cylinder car had become the 3000 — and was sold either as a two-seater or a 2+2.

It is now one of those well-known motor industry stories that Austin's chairman, Leonard Lord, first saw the Healey 100 prototype at the 1952 Earls Court Motor Show, did an immediate deal with Donald Healey, and the Austin-Healey marque was created. The result was a range of ruggedly attractive sports cars, which were persistently improved over the years.

Originally, body chassis units were produced by Jensen Motors, at West Bromwich, with final assembly and insertion of Austin engines and running gear at the BMC factory at Longbridge, but from the autumn of 1957 final assembly was moved to BMC's MG factory at Abingdon, near Oxford.

The first cars had 2.6 litre/90 bhp four-cylinder units (the ex-Austin A90 design, in fact), and three-speed gearboxes with overdrive, but within three years a four-speed gearbox had been provided, and tuned-up 100M (110 bhp) or 100S (132 bhp, with a light-alloy body) versions had appeared.

From 1956 there was a two inch longer wheelbase, 2+2 seating, and BMC's six-cylinder engine. This grew up to 2.9 litres/124 bhp in 1959, thus creating the Austin-Healey 3000, a model which was progressively made faster and more powerful. A wrap-round screen and wind-up windows were fitted from mid-1962.

AUSTIN-HEALEY SPRITE/MG MIDGET

Produced At Abingdon, UK, 1958 to 1971 (Sprite) and 1961-1979 (Midget). Production: 48,999 'Frog-eye' Sprites, 80,360 other Sprites, 226,526 Midgets. **Data for 1,275cc model** Unit-construction pressed-steel body/chassis unit. Front-mounted ohv four-cylinder engine, driving rear wheels. 1,275cc, 65 bhp at 6,000 rpm. Four-speed gearbox. Coil spring independent front, leaf-spring live-axle rear suspension; rack-and-pinion steering; front discs, rear drum brakes; 5.20-13in tyres. Wheelbase 6ft 8in; length 11ft 5.5in; width 4ft 5in. Unladen weight 1,720 lb. **Derivatives** 'Frog-eye' Austin-Healey Sprite was the original 948cc/43 bhp model, with drum brakes. From 1961 Sprite was re-styled, and badge-engineered MG Midget joined it. 1,098cc from late 1962, 1,275cc from late 1966, and different (Triumph) 1,493cc engine from late 1974. Sprites in 1971 were only badged as 'Austin'. **Performance potential** 1,275cc: Top speed 94 mph; 0-60mph 14.1 sec; standing ¼-mile 19.6sec; typical fuel consumption 33 mpg (Imperial).

Healey's liaison with BMC over the Austin-Healey 100 sports car project was a great success, so Donald Healey was then invited to produce a further design using A35 and Morris Minor components. The result was the Austin-Healey Sprite of 1958, a car now universally called the 'Frog-eye' due to the distinctive styling, with headlamps partly recessed into the bonnet panel.

This was a perky little sports car, with rather twitchy handling

The first Sprite was nicknamed 'Frog-eye' for very obvious reasons!

characteristics due to the layout and location of the rear suspension, but it had a great deal of character and sold well in the United States. There was no denying, however, that its looks were against it, and in 1961 it was restyled on more conventional lines, complete with an opening boot lid, and henceforth was sold as an Austin-Healey Sprite or an MG Midget, there being virtually no difference between the two cars.

For the next decade the car changed only gradually — a new engine tune one season, a better rear suspension another, with front wheel disc brakes (1962) and wind-up windows (1964) introduced to keep the car abreast of trends and competitive with the Triumph Spitfire — but the basic styling remained the same. Every 'Spridget' had a rather cramped interior, but a very sporting character and excellent steering, which made it an ideal competition driving test machine. There was also a great deal of motor racing success with heavily-modified examples.

Austin-Healeys became Austins for 1971, as Donald Healey's royalty agreement ran out, and after this there were only MG Midgets. MG enthusiasts were horrified to see an engine transplant for 1975 — the 1.5-litre Triumph Spitfire unit being specified — though this gave the car a 100 mph top speed at last, even though the looks had been ruined by the fitting of vast black plastic bumpers. Very little improvement was made to the car after that, and the last was built in 1979, more than 21 years after the original 'Frog-eye' had been introduced.

By the mid-1960s, Sprites had more conventional styling, and wind-up door windows.

BMW 507

Produced At Munich, West Germany, 1956 to 1959. Production: 253 cars. **Data** Tubular chassis, with steel body shell. Front-mounted ohv V8 engine, driving rear wheels. 3,168cc, 150 bhp at 5,000 rpm. Four-speed gearbox. Torsion bar independent front, torsion bar live axle rear suspension; bevel gear steering; four-wheel drum brakes (final few with front discs); 6.00-16in tyres. Wheelbase 8ft 1.6in; length 14ft 4.4in; width 5ft 5in. Unladen weight 2,935 lb. **Derivatives** None, though long-wheelbase 503 was mechanically related. **Performance potential** Top speed 124 mph; 0-60mph 8.8sec; standing ¼-mile 16.5 sec; typical fuel consumption 20 mpg (Imperial).

It was amazing that BMW could even contemplate building the very expensive, very exclusive Type 507 as they had struggled to become re-established after losing everything in the Second World War. The car was a beautifully sleek two-seater, based around a short-wheelbase version of the Type 502's running gear, complete with the 3.2 litre V8 engine. Styling was by Count Albrecht Goertz, and the car certainly had performance to match its looks. Nowadays, though, the saloon car character of its chassis shines through.

The 507 was a very limited production machine, as you would guess by studying details of its construction. It was closely related, mechanically, to the longer-wheelbase Type 503 Cabriolet, both of which were really superseded by the Bertone-styled 3200CS.

The BMW 507 was a status car for the Munich company, but not a profit-maker.

CHEVROLET CORVETTE

Chevrolet Corvette (1953 model) — **Produced** St Louis, Missouri, USA. Production: 300 cars in 1953 model year. **Data** Box-section separate chassis frame, with separate glass-fibre body shell. Front-mounted ohv six-cylinder engine, driving rear wheels. 3,857cc, 150 bhp at 4,200 rpm. Two-speed automatic transmission. Coil spring independent front, leaf-spring live axle rear suspension; worm-type steering; four-wheel drum brakes; 6.70-15in tyres. Wheelbase 8ft 6.4in; length 14ft 1.3in; width 6ft 0in. Unladen weight 2,715 lb. **Derivatives** 4,341cc V8 engine optional for 1955 models, 195 bhp at 5,000 rpm. Complete restyle for 1956. **Performance potential** 1953 model, 150 bhp: Top speed 107 mph; 0-60mph 11.0 sec; standing ¼-mile 18.0sec; typical fuel consumption 22 mpg (Imperial).

Chevrolet Corvette (1968 model Sting Ray, new style) — **Produced** St Louis, Missouri, USA. Production: 28,566 in 1968 model year. **Data** Box-section separate chassis frame, with separate glass-fibre body shell. Front-mounted ohv V8 engines (several size/power choices), driving rear wheels. From 5,357cc/300 bhp to 6,994cc/435 bhp. Three-speed/four-speed manual gearboxes/three-speed automatic transmission to choice. Coil spring independent front, transverse leaf/wishbone independent rear suspension; worm-type steering; four-wheel drum brakes; various 15in tyres. Wheelbase 8ft 2in; length 15ft 2in; width 5ft 9.3in. Unladen weight, from 3,055 lb. **Derivatives** Coupé and convertible styles — this basic model built, with many annual changes, to 1983. **Performance potential** 1968 model, with 350 bhp/5.7 litre tune: Top speed 128 mph; 0-60mph 7.7 sec; standing ¼-mile 15.6sec; typical fuel consumption 16 mpg (Imperial).

The American motor industry has rarely gone in for making out-and-out sports cars, which is why the arrival of the Chevrolet Corvette in 1953 was such a surprise. At the time, and still today, it offers a unique North American blend of style, performance and brash character.

Chevrolet's masters, General Motors, rarely took big risks, so the original Corvette had a specially designed chassis frame and a new body style in glass-fibre, but picked up all its running gear — engine, transmission, and suspensions — from existing Chevrolet touring models. In particular, the engine was the latest development of the legendary 'Stove Bolt' six, tuned to produce 150 bhp, and the standard transmission was a two-speed automatic.

The first Corvette, in other words, was not the ball of fire that it was eventually to become, and indeed it sold very slowly at first — only 300 cars in the 1953 model year, 3,640 in 1954 and 676 in 1955. This was due partly to the rather anonymous styling, partly to the limited performance, and partly to the US public's suspicion of a sports car built by Chevrolet, who had never attempted any such thing before.

The great surge forward, however, came in the 1955 model year (as it did for all Chevrolets) with the arrival of the new 4.3 litre V8 engine of 195 bhp. A year later that power increased to 225 bhp in one optional instance, and a three-speed manual transmission was also standardized. For the next 15 years, until the onset of exhaust emission

Above *The 1953 Corvette had very rounded lines, a wrap-round screen, and a six-cylinder engine.*

Below *Second-generation Corvettes looked more purposeful. This was a '61 model, with four headlamps.*

Bottom *Third-generation 'Vettes had an all-independent chassis, startling lines, and the name of 'Sting Ray'.*

This Corvette style, in essence, lasted from 1967 to 1983, and sold in huge numbers. This is a 1975 model.

regulations began to take its toll, each model year saw Corvettes getting faster, with better specifications.

The original body style, on an 8ft 6.4in wheelbase, was replaced by a larger and more purposeful two-seater on the same chassis for 1956; the same basic chassis, and body style, was retained until 1962, although engine power continued to increase, and the nose was given four headlamps in 1958.

The first big change came for 1963, with the launch of the 98in wheelbase, all-independent suspension Sting Ray, which had the famous 'shark-nose' and fastback coupé styling. On this car four-wheel disc brakes were standardized for 1965.

For 1968 the completely new 'Mako Shark' body style was launched, for the existing chassis, and this lived successfully on, in gradually evolving form, until 1983. An enormous number of V8 engine and transmission options, not to mention trim, decoration, and 'Special Edition' packs, were offered during this time, there being open, hardtop, fastback and hatchback derivatives, not to mention new noses and new tails, plus the compulsory energy absorbing bumpers from 1973. At the height of its career this particular type of Stingray was selling at the rate of 50,000 a year.

Then, in 1983, came the all-new, fifth-generation Corvette, still with all-independent suspension and four-wheel disc brakes, but with a new backbone chassis, and startlingly smooth styling which included a hatchback and removable roof panels. No doubt it will be destined for a long and successful career like all the other Corvettes — for Chevrolet's faith in this project has never been matched, let alone beaten, by any other American car maker.

CISITALIA 1100

Produced Turin, Italy, 1946 to 1948. Production: not known. **Data** Multi-tubular chassis frame, with separate light-alloy and steel body shell, various styles. Front-mounted ohv four-cylinder engine, driving rear wheels. 1,089cc, 50 bhp at 5,500 rpm. Four-speed gearbox. Transverse leaf and wishbone independent front suspension, leaf spring live axle rear suspension; worm and sector steering; four-wheel drum brakes; 5.00-15in tyres. Wheelbase 7ft 10.5in; length approx 12ft 0in; weight 1,425 lb. **Derivatives** Various body styles on same chassis. **Performance potential** Top speed (manufacturer's claim) up to 106 mph; typical fuel consumption 20 mpg (Imperial).

The short-lived Cisitalia project is not so much important for its engineering as it is for its styling. Under the skin was a multi-tube chassis and rather humble modified Fiat 1100 running gear. The bodies, however, were usually by Pininfarina, and their shapes really pointed the way to all the very best 1950s styles.

Cisitalia was Italian industrialist Piero Dusio's brainchild, with design by Dante Giacosa (who later became Fiat's technical chief). Only a limited number were built and —frankly — they were neither very refined, nor outstandingly fast, but they had a great deal of Italian character. That styling, however, will live for ever.

The Cisitalia, styled by Pininfarina, was the fashion predecessor of many later, famous coupés.

Above *Classic British style and engineering, plus brute US horsepower made the AC Cobra a unique 1960s proposition.*

Below *The running gear was used in many other 1950's Alfas, but the styling of the Giulietta Sprint was delicate, unique and pure Pininfarina.*

Above *The 'Big Healey', as it was affectionately known, was built in many forms — this was the 2.6 litre 100 Six.*

Below *For many years the Corvette was North America's* only *two-seater sports car, selling better and better as the years passed by.*

Top *The front-engined Ferrari Daytona was the world's fastest car in the late 1960s, though the mid-engined Boxer ran it close in the early 1970s. Both are already legendary.*

Above *The sensational Jaguar E-Type was a direct descendant of the racing D-Type of the 1950s. Doesn't it show?*

Below *Three generations of Jaguar XK all sold well, and went very quickly indeed — this was the XK150, introduced in 1957.*

Above *Perhaps the sexiest Supercar of all time — the Bertone-styled, V12, mid-engined Lamborghini Countach.*

Below *The famous Lamborghini Miura set so many standards — it was the first-ever mid-engined Supercar, and that was a transverse V12 unit!*

DAIMLER SP250

Produced Coventry, UK, 1959 to 1964. Production: 2,648 cars. **Data** Box-section separate chassis frame, with glass-fibre body shell. Front-mounted ohv V8 engine, driving rear wheels. 2,548cc, 140 bhp at 5,800 rpm. Four-speed gearbox, or three-speed automatic transmission. Coil spring independent front, leaf-spring live axle rear suspension; cam-and-lever steering; four-wheel disc brakes; 5.90-15in tyres. Wheelbase 7ft 8in; length 13ft 4.5in; width 5ft 0.5in. Unladen weight 2,220 lb. **Derivatives** Open sports and hard top styles; A,B, and C specifications as development progressed. **Performance potential** Top speed 121 mph; 0-60mph 10.2 sec; standing ¼-mile 17.8sec; typical fuel consumption 30 mpg (Imperial).

One of Coventry's oldest-established car companies, Daimler suffered several management and design upheavals during the 1950s, one of which resulted in the birth of the SP250 sports car. Called the 'Dart' at first (until Dodge objected ...), not only did it have strange styling and a glass-fibre body shell, but a magnificent new V8 engine.

The chassis was almost a straight copy of the Triumph TR3 layout, as was the gearbox, but there was a longer wheelbase, and the SP250 was also faster than the Triumph. At first the build quality was somewhat doubtful, but after Daimler were taken over by Jaguar matters improved.

Sales in the USA were never as high as hoped (it was, after all, the beginning of the E-Type era!), even though an automatic transmission was available. Jaguar-Daimler considered a re-style but abandoned it — and no further Daimler sports car was ever built.

The SP250 went much better than it looked, though the styling still has some devotees.

DATSUN 240Z FAMILY

Produced In Japan, 1969 to 1978. **Data for 240Z** Unit-construction pressed-steel body/chassis assembly. Front-mounted ohc six-cylinder engine, driving rear wheels. 2,393cc, 151 bhp (gross) at 5,600 rpm. Four-speed or five-speed gearbox. Coil spring/MacPherson strut independent front and rear suspensions; rack-and-pinion steering; front disc, rear drum brakes; 175-14in tyres. Wheelbase 7ft 6.7in; length 13ft 6.8in; width 5ft 4.1in. Unladen weight 2,300 lb. **Derivatives** 260Z, with 2,565cc/162 bhp (gross) from 1973, and 260Z 2+2, with 12in longer wheelbase and 2+2 seating. 280ZX of 1978 was completely different monocoque, with different suspension. **Performance potential** 240Z: Top speed 125 mph; 0-60mph 8.0sec; standing ¼-mile 15.8 sec; typical fuel consumption 24 mpg (Imperial).

Until the 1960s Datsun produced a series of boringly styled cars, so the arrival of the now-famous Z-Car in 1969 was a real surprise. The company had produced rather 'dumpy' sports cars before this, but the sleek 240Z was a revelation. Until it was displaced by the second-generation 280ZX in 1978, a somewhat less attractive machine aimed directly at the North American market, the Z-Car broke sales records *and* set styling standards.

The original shape was by Count Albrecht Goertz, whose earlier work had included the BMW 507, and its long nose hid the familiar single-cam Datsun six-cylinder engine. The 240Z was purely a two-seater, and had all-independent suspension, allied to rack and pinion steering. In character, but certainly not in looks, the 240Z was really a latter-day Austin-Healey 3000 — some say that Datsun had studied this car carefully before tooling up for their own coupé.

Although the 240Z was an instant success — in motor sport as well as in the showrooms — Datsun immediately set about improving it. There were 2 litre and twin-cam six-cylinder versions, never sold outside Japan, but the first major change for the outside world was the announcement of the 260Z of 1973, which was really a 240Z with the larger and theoretically more powerful 2.6 litre engine. Road tests, however, showed that the new car was actually slower than the original!

Then, at about the same time, Datsun also revealed the 260Z 2+2, which looked almost the same as the two-seater, but had 2+2 seating in a 12 inch longer wheelbase. There was no open-topped version of the design — but a Targa style eventually appeared on the next generation, the 280ZX.

The Z-Car sold amazingly well in the USA, where it was also available with automatic transmission as an option: the success was undoubtedly due to a combination of attractive styling, high performance, and Japanese reliability and after-sales service. The 280ZX which replaced it, in 1978, never had such pure-bred appeal to enthusiasts.

Above *The Z-Car was often customised by its keen owners — the sun-roof and the wheels on this car are non-standard.*

Below *With careful restyling, and a longer wheelbase, the 260Z 2+2 was still a beautiful car.*

FERRARI INTER FAMILY

Produced At Maranello, Italy, 1947 to 1953. Production: 38 166 Inters, 25 195 Inters, 80 212 Inters. **Data for 212 Export** Tubular chassis, with separate steel and alloy body shell. Front-mounted ohc V12 engine, driving rear wheels. 2,562cc, 140bhp/170 bhp at 5,600/6,500 rpm. Four-speed gearbox. Transverse leaf spring and wishbone independent front, leaf spring live axle rear suspension; worm and sector steering; four-wheel drum brakes; 5.90-15in tyres. Wheelbase 7ft 4.6in; overall length 12ft 2.5in; width 5ft 1.4in. Unladen weight 2,035 lb, depending on body style. **Derivatives** 166 Inter was original, with 1,995cc/110 bhp, 195 Inter followed in 1951 with 2,340cc/130bhp, while 212 Inter had 140 bhp 2,562cc engine. 250 Europa (2,963cc/200 bhp upwards) was developed from this range. **Performance potential** 212 Export/140 bhp: Top speed 116 mph; 0-60mph 10.5 sec; standing ¼-mile 17.5 sec; typical fuel consumption 15 mpg (Imperial).

Enzo Ferrari had already established an illustrious career — as racing driver, then as racing team manager, with Alfa Romeo — before setting

The first Ferrari road car was the 166 Inter, with smooth styling and the famous V12 engine.

The 195 Ferrari Inter coupé, with 2.3 litre engine, already had that obvious Italian style.

up his own manufacturing business in the 1940s. The first Ferrari road cars were the Inter models, important not only for that reason but also because they were the first to use the famous Colombo-designed V12 engines.

The basis of all early Ferrari road cars was a simple tubular chassis frame, with transverse leaf spring independent front suspension and half-elliptic leaf spring suspension of the rear live axle. The magnificent engine and the four-speed transmission were of Ferrari's own manufacture, though then —as now — he bought the cars' coachwork from outside specialists. Except that the engines were de-tuned for road use, the Inters were very closely related indeed to the racing Ferraris of the period.

The engine itself, which had wet cylinder liners, started life as a 1.5 litre racing unit, but the first road engine was the Type 166 2.0 litre. Over the years, to the 1960s, and with hundreds of modifications, this engine proved capable of expansion to 3.3 litres, giving well over 300 bhp. In the 166 Inter, 2.0 litres, single overhead camshafts and 110 bhp was enough.

All the Inters of the 1947-53 period had the same urgent character, and all had two-seater coachwork. In their sound, their performance, and their looks they were unmistakably Ferrari — and the lasting legend of the Ferrari road car was soon established. In those days, performance and engine quality were more highly regarded than to almost everything else, for roadholding and anti-corrosion protection were not considered to be as important.

There was little overlapping of models — the 166 being replaced by the 195 in 1950, and the 212 in 1951, while the 212 Export had a shorter wheelbase. As can be seen from the production figures, a Ferrari Inter was rare at the time, and is even more rare and prized today.

FERRARI AMERICA AND SUPER AMERICA

Produced At Maranello, Italy, 1951 to 1959. Production: 22,340 Americas, six 342 Americas, 13,375 Americas, and 38,410 Super Americas. **Data for Type 410 Super America** Tubular chassis, with separate steel and alloy body shell. Front-mounted ohc V12 engine, driving rear wheels. 4,961cc, 340 bhp/400 bhp at 6,000/6,500 rpm. Four-speed gearbox. Coil spring independent front, leaf spring live axle rear suspension; worm and roller steering; four-wheel drum brakes; 5.50/6.50-16in tyres. Wheelbase 8ft 6.4in; length 15ft 5in; width 5ft 6.5in. Unladen weight 2,900 lb — all depending on individual body style. **Derivatives** 340 America was original, with 4,102cc/200 bhp, 342 America was an evolution, 375 America had 4,523cc/300 bhp on longer wheelbase, while 410 Super Americas all had 4,961cc V12 engine. **Performance potential** 1959 400 bhp Type 410: Top speed 165 mph; 0-60mph 6.6sec; standing ¼-mile 14.6 sec; typical fuel consumption 14 mpg (Imperial).

All the earlier Ferraris were exciting, and to some extent hand-built, but some were even more exclusive than others. To afford a 1950s Ferrari was one thing, but to rise to the heights of one of the 'America' family was a very special achievement indeed! A study of numbers built — only 79 cars in the 1950s — confirms this.

All of them used the same basic type of 'early-Ferrari' layout, all had the largest current type of Ferrari V12 engine and all were fitted with

Grand Prix driver De Portago with a Type 410 built in 1956.

This magnificent Type 410 Super America was built in 1957.

specially-commissioned body styles. The first cars were the 340-series Americas, with transverse leaf independent front suspension and 220 bhp 4.1 litre engines (that sounds puny today, but was considered exceptional at the time). Their running gear was all closely related to the Type 340 Mexico and 340MM sports racing machines of the day. In spite of the size and weight of the America models, they were only two-seaters.

Ferrari's problem was not that the series was exclusive and expensive (there were sufficient wealthy customers to buy all the cars he wanted to build), but that at first an America was no faster than several other far cheaper American-engined rivals being sold in the USA. Accordingly, the story of the next few years was a persistent upgrading of the 'America' theme to make the cars faster and ever more impressive than before.

The 340's successor was the 342 of 1952, but in 1953 this car was replaced by the 375 America, which used the 250 Europa's chassis, had 4.5 litres and 300 bhp and a top speed claimed to be 155 mph. Then came the Type 410 Super America of 1956, a much better car with coil spring independent front suspension and 340 bhp from its 4.9 litre engine, which was the absolute limit of 'stretch' for the original 'Lampredi' V12 unit. The third series of Type 410s was even faster, with 360 bhp, and this was nearly the end of the road for this particular strain.

There were more Super Americas and Superfasts in the 1960s, but these were different in many ways and with different V12 engines. Production of the famous 250 Series was booming by this time and the day of the bespoke, hand-crafted Ferrari road car was nearly over.

FERRARI 250GT FAMILY

Ferrari 250GT Berlinetta — **Produced** At Maranello, Italy, 1959 to 1962. Production: 2,500 of all 250GT models, of which 155 were short-wheelbase Berlinettas. **Data for SWB 250GT Berlinetta** Tubular chassis, with steel and alloy body shell. Front-mounted ohc V12 engine, driving rear wheels. 2,953cc, 220/240 bhp at 7,000 rpm. Four-speed gearbox. Coil spring independent front, leaf spring live-axle rear suspension; worm and sector steering; four-wheel disc brakes; 185-15in tyres. Wheelbase 7ft 10.5in; length 13ft 8in; width 5ft 6in. Unladen weight 2,600 lb (competition cars 2,400 lb). **Derivatives** 250GT Europa was the original type, followed by Boano styles, and Pininfarina coupés and spyders from 1958. **Performance potential** 250GT road car: Top speed 150 mph; 0-60mph 9.2sec; standing ¼-mile 17.0sec; typical fuel consumption 15 mpg (Imperial).

Ferrari 250GT Berlinetta Lusso — **Produced** At Maranello, Italy, 1962 to 1964. Production: 350 Lussos. **Data** Tubular chassis, with steel body shell. Front-mounted ohv V12 engine, driving rear wheels. 2,953cc, 240/250 bhp at 7,500 rpm. Four-speed gearbox. Coil spring independent front, leaf spring live-axle rear suspension; worm and sector steering; four-wheel disc brakes; 185-15in tyres. Wheelbase 7ft 10.5in; length 14ft 5.6in; width 5ft 9in. Unladen weight 2,890 lb. **Derivatives** None. **Performance potential** Top speed 142 mph.

When considering the Ferrari '250' family the problem is always knowing where to start and where to stop, for some were road cars and some were racing GT cars. Not all were 250GTs either...

In 1957 Boano were producing these two-door coupés on the Ferrari 250GT chassis.

This is one of the early (1957-58) Pininfarina Cabriolets on the 250GT chassis.

The very first Ferrari 250 was the Europa model of 1953, but this was a very limited-production (17-off) derivative of the Type 375 America, with a smaller-capacity version of the massive 'Lampredi' V12 engine. All other 250GTs, which began a long and illustrious career in 1954, used one or other version of the original 'Colombo' V12, of 2,953cc.

Between 1954 and 1964, when the last 250GT-type car was built, Ferrari's annual production of road cars rocketed from 35 to 670 a year, all this expansion being to the credit of these cars. *That* is a measure of how important the 250GT was, and is, to the Ferrari company's standing.

Over the years there were many mechanical derivatives, but the basis of every 250GT was a simple but rugged tubular chassis frame, with coil spring independent front suspension, and a beam rear axle located by radius arms and sprung on old-fashioned 'cart' springs. Every type had the single-overhead-cam-per-bank 'Colombo' V12 engine, the power of which was progressively up-rated over the years. At first the cars had a four-speed transmission, to which an overdrive was later added.

Most of them were two-seaters — open or closed — but from 1961 there was also the 250GTE (or 250GT 2+2) in which two extra seats were inserted after the engine and bulkhead had been moved as far forward as possible.

The 250GT Europa of 1954-55 had a 102.4in wheelbase, and was followed in 1956 by the 250GT 'Boano' and 'Ellena' styles, built by the Turin-based coachbuilders of that name on the same chassis. Pininfarina also built some stunningly beautiful cabriolets from 1957, and in 1958 his new coupé design took over from 'Ellena' with electrically-operated overdrive as an early modification and four-wheel disc brakes

The most famous 250GTs of all were the SWB (short wheelbase) Berlinettas, which raced so successfully in the early 1960s.

from the end of 1959. The Spyder California superseded the Pininfarina cabriolet in 1958, being a more sporty car and bodied for Ferrari by Scaglietti in Modena: short-wheelbase versions of this car were successful in competition. Then came the second-series Pininfarina Cabriolets of 1959, which were rather sumptuous open-topped machines.

Two even more carefully developed machines followed for the 1960s. One was the 250GTE (or 250GT 2+2), which had a more spacious cabin but frankly very little rear seat leg-room, while the other was the sinuous and quite stunning 250GT Berlinetta Lusso of 1962. Both were built until the new all-independent suspension cars arrived in 1964.

The 2+2 was long considered, rather disparagingly, as not a very sporting Ferrari, though it still had formidable performance and made all the right sort of Ferrari engine noises. Consequently it has not survived in such large numbers to the 'classic conscious' 1980s.

The Lusso, on the other hand, was always a much-loved machine. Whereas the 2+2 had rather angular lines, the Lusso had a sweeping style, probably unsurpassed up to that time by Pininfarina (though built by the Scaglietti works, as usual). It had a standard-specification 240 bhp engine on the shorter (94.5in) wheelbase, and a top speed of more than 140 mph was attainable. No fewer than 350 of these delightful coupés were built in two years.

Even more exciting 250GTs, however, were the short wheelbase coupes, known as 250GT SWBs, which rode on the 94.5in wheelbase and were mostly used in competitions with light-alloy bodies, though steel-bodied versions were also employed as truly exhilarating road machines. The normal engine tune gave 240 bhp, but competition machines had up to 280 bhp and top speeds of over 150 mph. Only 155 of these cars were built between 1959 and 1962. Lastly, and most amazing of all, there were the 250GTOs, which were competition cars, pure and simple, in which the peak power went up to 300 bhp. There was a five-speed gearbox and the body style was longer, lower and lighter than before. Only 42 GTOs were built, these nowadays being the most desirable '250GTs' of all.

In total, about 2,500 of the various 250GT types were built, making them by far the most numerous Ferraris made up to that time.

The most elegant 250GT of all, without doubt, was the sinuously-styled Lusso of 1962-64.

FERRARI 275GTB AND DAYTONA FAMILY

Ferrari 275GTB family — Produced At Maranello, Italy, 1964 to 1968. Production: 954 275GTB/275GTS coupés/spyders. **Data for 275GTB** Tubular chassis frame, with steel and light-alloy separate body shell. Front-mounted ohc V12 engine, driving rear wheels. 3,286cc, 280 bhp at 7,500 rpm. Five-speed, rear-mounted gearbox. All-independent coil spring/wishbone suspension, front and rear; worm and sector steering; four-wheel disc brakes. 185VR-14in tyres. Wheelbase 7ft 10.5in; length 14ft 6in; width 5ft 7in. Unladen weight 2,495 lb. **Derivatives** 275GTB/4 had twin-overhead-cam per bank engine, producing 300 bhp. 330GTS was 4.0 litre-engined version of 275GTS, and 365GTS had 4.4 litres. **Performance potential** 275GTB: Top speed 153 mph; 0-60mph 6.0sec; standing ¼-mile 14.0sec; typical fuel consumption 18 mpg (Imperial).

Ferrari Daytona family — Produced At Maranello, Italy, 1968 to 1974. Production: 1,412 Daytonas, coupés and spyders. **Data for Daytona** Tubular chassis frame, with separate steel and light-alloy body shell. Front-mounted 2 ohc V12 engine, driving rear wheels. 4,390cc, 362 bhp at 7,500 rpm. Five-speed rear-mounted gearbox. All-independent coil spring/wishbone suspension, front and rear; worm and nut steering; four-wheel disc brakes. 215VR-15in tyres. Wheelbase 7ft 10.5in; length 14ft 6in; width 5ft 9.25in. Unladen weight 3,530 lb. **Derivatives** Coupé and Spyder versions available. **Performance potential** Top speed 174 mph; 0-60mph 5.4sec; standing ¼-mile 13.7sec; typical fuel consumption 14 mpg (Imperial).

Until the early 1960s every Ferrari road car used the same type of chassis frame, with large-diameter tubes, and a heavy live beam rear axle suspended on half-elliptic leaf springs. By this time a change to a more modern layout was overdue and the arrival of the 275GTB in 1964 was therefore very important indeed to Ferrari's future prospects.

The 275GTB line was on sale for four years and the now-legendary Daytona style which followed was built until 1973. These were the true two-seater Supercars, but several other models evolved from the same basic layout, including the 330GTC range, while the largest-engined Ferraris also took on board similar suspension systems.

Central to the new design generation was the adoption of independent suspension to front *and* rear wheels, and on the ultra-sporting 275GTB there was a new five-speed gearbox in unit with the chassis-mounted final drive. On the first 275GTBs produced the engine and the rear-mounted transaxle were linked by a slim solid propeller shaft with a centre steady bearing, but this was changed about a year later to incorporate a solid torque tube shaft linking the two masses, and enclosing the propeller shaft itself.

The new multi-tube chassis had what, for Ferrari, was a traditional 94.5 in wheelbase, and naturally it had a 60-degree V12 engine. This was a 3.3 litre version of the famous old 'Colombo' design, much-modified by

Above *In 1964 Ferrari produced the 275GTB model, their first to have all-independent suspension.*

Below *275GTBs were not meant to be racing cars, but many owners used them successfully in competition.*

this time, of course, and it produced 280 bhp. There were two body styles — a coupé with headlamps hidden behind perspex cowls, and a cut-off tail, built by Scaglietti, and the 275GTS Spyder built by Pininfarina.

After only two years Ferrari made a great car even better by fitting twin-cam cylinder heads to the same basic engine, which helped the peak power rise to 300 bhp, and the model name became 275GTB/4. Next along, and so typical of the way that Ferrari could stir up all his available hardware and produce intriguing new combinations, were the 330GTC (a two-door two-seater coupé) and the 330GTS (which looked the same as the 275GTS Spyder). In each case the rear-transmission, all-independent suspension chassis was used, but the engine was a large 'Lampredi-type' 4 litre with 300 bhp.

There was more to come. In the autumn of 1968 Ferrari replaced the successful 275GTB/4 with an even more startling two-seater Supercar, the fabulous 365GTB/4 Daytona coupé. This, without doubt, was then the fastest Ferrari to date, for independent tests put its top speed at 174 mph, with acceleration to match.

The same efficient chassis frame as the 275GTB/4 was retained, but the engine was a large 4.4 litre V12, with twin overhead camshaft cylinder heads and no less than 352 bhp. The styling, as usual by Pininfarina, but built by Scaglietti, was even more graceful than the

Ferrari's magnificent 362 bhp Daytona was the world's fastest car in the late 1960s. You can see why!

Sister car to the 275GTB was the 275GTS Spider — same chassis, but completely different body style.

deposed 275GTB/4, with a low nose, faired-in headlamps and a long sloping tail: from 1971 all Daytonas had headlamps covered by flaps. A 365GTS/4 Spyder followed in 1969.

The same engine size — actually 4,390cc — was also picked up by the 365GTC and 365GTS models of 1968-69, which used the same chassis and basic styling of the 330GTC/330GTS cars which they replaced. However, when reviewing the lineage and evolution of modern Ferraris, it is always wise to hedge one's bets, as all to some extent can be classed as Supercars and therefore classic sporting cars in their own right. The 365GT 2+2 of 1967-71 was a very large four-seater coupé, yet it used a similar all-independent suspension chassis to the 275GTB but of 8ft 8.3in wheelbase, with a 4.4 litre single-overhead-cam engine and a front-mounted gearbox. On the other hand, the 365GTC/4 of 1971 which replaced it had a shorter 8ft 2.4in wheelbase version of the same chassis, a 340 bhp engine with entirely different cylinder heads from the Daytona *or* the 365GT 2+2, the front-mounted gearbox and neat, almost Daytona-like, body style which hid 2+2 seating. To follow *that*, the full four-seater 365GT4 2+2 was very definitely a two-door saloon on an 8ft 10.3in wheelbase version of the chassis!

At Ferrari, for sure, nothing was ever simple or logical — but it was never boring either. No wonder the cars are so avidly studied by every motoring enthusiast.

FERRARI DINO 6-CYLINDER AND 8-CYLINDER FAMILY

Ferrari Dino 206/246 and 308 families — Produced at Maranello, Italy, 1967 to 1973 (V6 cars), 1973 to date (V8 cars). Production: 150 Type 206, 3,912 Type 246, and 10,000 Type 308s of all types by mid-1985. **Data for 246 GT Dino** Tubular chassis, with separate steel body welded to it on assembly. Mid-transversely-mounted 2ohc V6 engine, driving rear wheels. 2,418cc, 195 bhp at 7,600 rpm. Five-speed gearbox. All-independent coil spring/wishbone suspension; rack and pinion steering; four-wheel disc brakes. 205VR-14in tyres. Wheelbase 7ft 8.2in; length 13ft 9in; width 5ft 7in. Unladen weight 2,400 lb. **Derivatives** 206 was original, with 1,987cc/180 bhp, and alloy body. Coupé and spider versions of 246. 308 family used same wheelbase and chassis, but new 2,922cc V8 engine. Power originally 255 bhp. 2+2 Bertone and Pininfarina (called Mondial) styles. 2-seater GTB/GTS Pininfarina styles. Injection, and later 4-valves-per-cylinder heads, in early 1980s. **Performance potential** Dino 246GT: Top speed 148 mph; 0-60mph 7.1sec; standing ¼-mile 15.4 sec; typical fuel consumption 23 mpg (Imperial). 308GTB, original type: Top speed 154 mph; 0-60mph 6.5 sec; standing ¼-mile 14.8 sec; typical fuel consumption 21 mpg (Imperial).

When Enzo Ferrari's much-loved son Dino died at a tragically early age his father conferred his name on the family of new V6 engines then being developed. When that engine was productionized by Fiat in the mid-1960s it was almost inevitable that the first-ever mid-engined Ferrari, which used it, should also be called a Dino. Compared with all other Ferraris it was so very different that for a time the company tried to market it *without* the Ferrari name appearing anywhere on it.

The original car was the Dino 206, with an aluminium 2 litre

Ferrari Dino 246GTB and GTS looked virtually the same—this is the GTS, with removable soft-top.

engine/gearbox unit transversely mounted across the chassis behind the seats. The multi-tube chassis, all-independent suspension and Pininfarina-styled body were all 'typical-Ferrari' of the period.

Within two years of its launch, however, the Dino 206 became the Dino 246 — not only with the enlarged, iron-block, 2.4 litre V6 engine, but also with a slightly longer wheelbase and steel instead of light-alloy body panels. Although there were to be many more 246s than 206s, the earlier cars have withstood the passage of time much better and are more desirable today.

The Dino 246 coupé was joined by the soft-top GTS in 1972, but both were displaced by the new Type 308 models in 1973. After that, the old engine/transmission unit found a home in the Lancia Stratos, performing very successfully in major rallies throughout the world.

The first Dino 308 was the Bertone-styled 2+2-seater GT4, but this was quite overshadowed by the 308GTB two-seater, by Pininfarina, which came on the scene in 1975. Both used the same basic chassis as the six-cylinder car, but employed an all-new 90-degree 2.9 litre V8 unit.

Since then the range has proliferated and improved. The GTB was joined by the GTS Spider in 1977, and the GT4 was displaced by the 2+2-seater Mondial in 1981. From 1981 the engines gained fuel injection and for 1983 there were four-valves-per-cylinder heads. More recently, in 1984, the range was topped off by the 200-off turbocharged 400 bhp GTO, similar but by no means the same as the 308GTB under the skin!

For 1986 the engines were enlarged to 3.2 litres, and the model name became 328GTB.

The famous 308GTB of 1976, now no longer called the Dino, had a 2.9 litre V8 engine.

FERRARI BOXER AND TESTAROSSA

Produced At Maranello, Italy, 1973 to 1984 (Boxer), 1984 on (Testarossa). Production: 385 Type 365 Boxers, 1,936 Type 512 Boxers. Testarossa production proceeding. **Data for Boxer Type 365GT4/BB** Tubular chassis, with steel body welded to it on assembly. Mid-mounted 2 ohc flat-12 cylinder engine, driving rear wheels. 4,390cc, 380 bhp at 7,200 rpm. Five-speed gearbox. All-independent coil spring and wishbone suspension; rack and pinion steering; four-wheel disc brakes; 215VR-15in tyres. Wheelbase 8ft 2.4in; length 14ft 3.7in; width 5ft 10.9in. Unladen weight 2,725 lb. **Derivatives** 512BB followed 365BB in 1976, with 4,942cc/360 bhp, finally displaced by completely new-style Testarossa, on same chassis, with 4-valves-per-cylinder/390 bhp in 1984. **Performance potential** Boxer 365BB: Top speed 171 mph; 0-60mph 6.5sec; standing ¼-mile 14.0sec; typical fuel consumption 14 mpg (Imperial).

Although the first mid-engined Ferrari road car was the V6 Dino, the first 12 cylinder-engined Supercar was the Boxer, announced in 1971 and built until 1984. The Boxer, and its successor the Testarossa, are among the best of the best. One runs out of superlatives in describing them and their performance...

Ferrari's fastest road car at the end of the 1960s was the V12-engined Daytona, and a quite supreme new car would be needed to replace it properly. The 365GT4/BB — always known as the 'Boxer' because of its flat-12 engine layout — was that car. It combined everything Ferrari had so far learned from racing and road cars in one elegantly styled package.

The Ferrari 512iBB — with a 170 mph top speed potential. Need we say more?

For 1985 Ferrari up-dated the Boxer concept with this startling style, gave the engine 4-valve cylinder heads and called it the Testarossa.

There was the usual type of Ferrari multi-tube chassis frame, all-independent suspension and disc brakes, with styling by Pininfarina and body construction by Scaglietti, but the *pièce de résistance* was the engine/transmission layout. The engine was a new 4.4 litre four-cam flat-12 unit (actually with the same bore and stroke as the superseded Daytona), and was mid-mounted behind the two seats, above the complex new transmission which included a five-speed gearbox and the final drive unit.

Apart from the fact that the engine was perhaps a little higher than the ideal position, this was a state-of-the-art chassis in all respects and the roadholding certainly matched the straight-line performance.

The ever-tightening USA exhaust emission regulations made Ferrari enlarge the engine to 4.9 litres in 1976, actually with less power and a slightly lower top speed than before, but the 512BB was still one of the fastest cars in the world and sold strongly to 1984.

At this point it was replaced by the startlingly-styled new Testarossa, which had a Pininfarina shape over the same basic mid-engined chassis. The engine, however, now had four valves per cylinder and no less than 390 bhp. No wonder a top speed of more than 180 mph was claimed — and achieved!

FIAT 1200/1500 FAMILY

Produced In Turin, Italy, 1959 to 1966. Production: 11,851 1200s, 22,630 1500s/1500S models, and 3,089 1600Ss. **Data for 1200** Unit-construction pressed-steel body/chassis assembly. Front-mounted ohv four-cylinder engine, driving rear wheels. 1,221cc, 58 bhp at 5,300 rpm. Four-speed gearbox. Coil spring independent front, leaf spring live axle rear suspension; worm and roller steering; four-wheel drum brakes; 5.20-14in tyres. Wheelbase 7ft 8.1in; length 13ft 2.7in; width 4ft 11.9in. Unladen weight 1,994 lb. **Derivatives** 1500S was the same car, but with special OSCA-designed 2 ohc 1,491cc/80 bhp engine, mostly built with front disc brakes, 1500 was next-generation 1200 with different ohv 1,481cc/72 bhp engine while 1600S was 1,568cc/90 bhp 2 ohc model, with four-wheel discs from 1963, five-speed box from 1965. **Performance potential** 1200: Top speed 90 mph; 0-60 mph 19.1sec; standing ¼-mile 21.0 sec; typical fuel consumption 36 mpg (Imperial). 1500S Twin-Cam: Top speed 105 mph; 0-60 mph 10.6 sec; standing ¼-mile 18.5 sec; typical fuel consumption 30 mpg (Imperial).

Fiat's first attempt at an 1100-based sports car, the *Trasformabile*, was a styling disaster but the 1200-based cars which followed in 1959 were much more successful. It was an interesting combination of Pininfarina styling and little-modified Fiat saloon car running gear — except for the limited production OSCA-engined twin-cam, which made all the headlines.

From 1959 to 1966 there was one smart but unspectacular body style, which Pininfarina not only shaped but also produced in quantity. Underneath it was the modified 1200 saloon's structural floorpan and its front and rear suspensions. Most of the cars were drop-head Cabriolets, but there was also an optional hardtop for this shell, while Pininfarina themselves also produced limited numbers of a coupé which bore a striking resemblance to the Lancia Flavia Coupé of the day.

The most important feature of this design was its choice of engines. The vast majority of all these cars had a normal Fiat pushrod unit (the 58 bhp/1,221cc '1200' until early 1963 and the completely different Fiat

The simple, uncluttered lines of the Fiat 1200 Cabriolet, all based on the underpan of the 1200 saloon.

Top *A 1500 pushrod engine model in white faces up to the twin-cam 1600S (dark coloured) in 1964.*

Above *The 1965 model 1600S had a five-speed transmission.*

1500 72 bhp/1,481cc thereafter). These cars, frankly, were not very fast but the exciting versions had a special twin-cam engine which OSCA had designed for their own use, but which they had then persuaded Fiat to adopt, tool and produce for this new sports car.

The OSCA twin-cam had 80bhp from 1,491cc at first, and 90 bhp from 1,568cc by late 1962, and surprisingly it was never used in any other Fiat. It made all the right noises, looked good as well as going well and provided the 1500S/1600S models with top speeds approaching 110 mph.

In Italy at the time the Fiat was usually overshadowed in publicity and 'image' terms by the Alfa Romeo Giulietta, but it always outsold that car. The specification was persistently improved — notably by the fitment of disc brakes in the 1960s (on all four wheels on the 1600S) and by the use of a new five-speed gearbox on the 1600S from 1965.

These cars were replaced by the 124 Sport Spiders in 1966.

FIAT 124 SPIDER/COUPÉ FAMILY

Produced In Turin, Italy, 1966 to 1982. (Spider production continues, latterly, as the 'Pininfarina 124 Spider Europa'.) Production: 178,439 Spiders, all types, 279, 672 Coupés, all types, and 1,013 124 Spider Abarth Rallyes. **Data for 124 Spider 1.4 litre** Unit-construction pressed-steel body/chassis assembly. Front-mounted 2 ohc four-cylinder engine, driving rear wheels. 1,438cc, 90 bhp at 6000 rpm. Five-speed gearbox. Coil spring independent front, coil spring live axle rear suspension; worm and roller steering; four-wheel disc brakes; 165-13in tyres. Wheelbase 7ft 5.8in; length 13ft 0.4in; width 5ft 3.5in. Unladen weight 2,083 lb. **Derivatives** 1.6 litre Spider arrived 1969 (1,608cc/110 bhp), 1.8 litre version 1972 (1,756cc/118 bhp), 2.0-litre version 1978 (1,995cc/87 bhp in USA, not sold in Europe). Equivalent 124 Sport Coupés, with 7ft 11.2in wheelbase, 2+2 seating, built to 1975, but no 2 litre version for any market. Plus 1,756cc/128 bhp, all-independent suspension Spider Abarth Rallye, 1972 to 1975. **Performance potential** 124 Spider 1.4 litre: Top speed 106 mph; 0-60mph 11.9 sec; standing ¼-mile 18.3 sec; typical fuel consumption 30 mpg (Imperial). 124 Spider 1.6 litre: Top speed 112 mph, 0-60mph 12.2 sec; standing ¼-mile 18.6 sec; typical fuel consumption 28 mpg (Imperial).

Without any doubt, the family of 124-based sporting cars was Fiat's most successful in this market. The X1/9 might be a prettier car, but the various 124s outsold it handsomely. As with the 1200/1500s, the concept of using a Fiat saloon's running gear was straightforward enough, though the sourcing of bodies was not. At the time, perhaps, not even Fiat expected the shape of the new Spider still to be selling after twenty years!

The Fiat 124 sporting twins of 1966/67 were the Spider and the Coupé, the latter styled by Fiat itself.

By 1975 the 124 Sport Spider was being sold only in the USA, with these bulky bumpers.

The story starts with the 124 saloon of 1966 and with the twin-cam 125 engine which was to follow a year later. Neither had any links with the past and both were vitally important to Fiat's commercial future. Then, in the autumn of 1966, Fiat announced the 124 Sport Spider, which used a shortened 124 floorpan and suspension, a tuned-up version of the twin-cam engine (the 125 was still not announced at that stage), with styling and bodyshell construction by Pininfarina.

A few months later, the Sport Coupé was also announced — this one having the saloon-length floorpan and smart coupé styling by Fiat themselves. The two cars, between them, would sell strongly until 1975, when the Sport Coupé was dropped. Some markets took these cars with four-speed gearboxes, but the vast majority had the five-speeder (first seen in the OSCA 1600S of 1965).

Here were two typical Italian sporting cars, the Spider with 2+2 seating, the Coupé a genuine four-seater, both having bags of character, good looks and a great deal of performance. The pity of it was that they tended to rust away rather rapidly, too, this being one of Fiat's less endearing habits in the 1960s and 1970s.

The original 1.4 litre grew to 1.6 litres for 1970 and to 1.8 litres for the 1973 model year. Then, three years after the coupé's demise, the engine grew again to 2 litres, where it remains. From 1982 Fiat officially withdrew from the sports car business, handing over complete manufacture to Pininfarina who badged it themselves instead of it being a Fiat.

In addition, one must mention the 124 Abarth Rallye, an 'homologation special' which had independent rear suspension, lightweight panels and 128 bhp.

FIAT DINO SPIDER AND COUPÉ

Produced In Turin, Italy, and (from 1969) Maranello, Italy, 1966 to 1973. Production: 1,163 Spider and 3,670 Coupé 2.0 litres, 420 Spider and 2,398 Coupé 2.4 litres. **Data for Dino Coupé 2400** Unit-construction pressed steel body/chassis assembly. Front-mounted 2 ohc V6 engine, driving rear wheels. 2,418cc, 180 bhp at 6,600 rpm. Five-speed gearbox. Coil spring and wishbone independent front, coil spring and semi-trailing arm independent rear suspension; worm and roller steering; four-wheel disc brakes. 205VR-14in tyres. Wheelbase 8ft 4.4in; length 14ft 9.5in; width 5ft 6.8in. Unladen weight 3,042 lb. **Derivatives** Original cars were Spider (on 7ft 5.8in wheelbase) and Coupé, with 1,987cc/160 bhp, and leaf spring live rear axle suspension. From late 1969 2400 model took over, with an equivalent Spider. 2400 assembly at Maranello (Ferrari factory). **Performance potential** 2.0 litre: Top speed 127 mph; 0-60mph 8.1sec; standing ¼-mile 16.0sec; typical fuel consumption 20 mpg (Imperial). 2.4-litre: Top speed 130 mph; 0-60mph 8.7 sec; standing ¼-mile 16.1 sec; typical fuel consumption 20 mpg (Imperial).

The name Dino is a giveaway, for it obviously, and correctly, links this Fiat with the Ferrari of the same name. When a new 1½ litre racing Formula 2 was proposed for 1967, Ferrari planned to use his Dino V6 engines in new single-seater cars. Since the rules demanded that 500 such engines had to have been built into road cars, Ferrari persuaded Fiat not only to produce a model using the engine, but to make the engines as well!

As with the contemporary 124 Sport models, there were two distinctly different Dinos — the Spider with style and production by Pininfarina and the Coupé with style and production by Bertone. Both had structural underframes designed and provided by Fiat — the Coupé having a considerably longer wheelbase — but all shared the same running gear.

In spite of its high performance, the Dino was not a thoroughbred at

Fiat produced a four-cam V6 engine for Ferrari and installed it in Spider and Coupé Dinos of their own.

Pininfarina styled the Dino Spider — an all-time classic shape.

first, for the transmission was modified Fiat 2300 and the front suspension was lifted from the 125 saloon, while the beam rear axle had a somewhat unsatisfactory linkage. But no-one could complain about its character (the spine-tingling noise of the Fiat-redesigned Ferrari engine saw to that), and the necessary 500 cars were soon built to allow Ferrari to go motor racing.

In the autumn of 1969, however, Fiat announced radically re-designed Dinos, with unchanged styling. The alloy-blocked 2 litre engine was replaced by a cast-iron-blocked 2.4, the gearbox was a new five-speed German ZF unit, while there was new independent rear suspension lifted from the latest Fiat 130 executive saloon. The cars were little faster, but much better-developed than before, and they continued to sell remarkably well for machines carrying such high price tags.

Fiat had taken a 50 per cent stake in Ferrari in 1969, and one result was that Dino assembly was concentrated thereafter at the Ferrari Maranello factory.

The four-cam V6 engines, incidentally, were also used in Ferrari's own Dino 206 and 246 models, and in the Lancia Stratos 'homologation special' — both transversely mid-mounted, with their own special transmission.

FIAT X1/9

Produced In Turin, Italy, 1972 to 1982. (Since 1982, badged as a 'Bertone' Fiat X1/9, assembled by Bertone.) Production: Approx 100,000 1300s, approx 50,000 1500s. **Data for X1/9 1300** Unit-construction pressed-steel body/chassis assembly. Mid, transversely-mounted, ohc four-cylinder engine, driving rear wheels. 1,290cc, 75 bhp at 6,000 rpm. Four-speed gearbox. All-independent suspension by coil springs and MacPherson struts; rack and pinion steering; four-wheel disc brakes. 165SR-13in tyres. Wheelbase 7ft 2.75in; length 12ft 6.75in; width 5ft 1.75in. Unladen weight 2,010 lb. **Derivatives** From late 1978 the X1/9 1500 took over, with 1,498cc/85 bhp, and five-speed gearbox. Bertone manufactured from 1982. **Performance potential** X1/9 1300: Top speed 99 mph; 0-60mph 12.7sec; standing ¼-mile 18.8 sec; typical fuel consumption 32 mpg (Imperial).

The X1/9 was not the first mid-engined sports coupé, but it was certainly the first to sell in large numbers at a relatively low price. Announced in 1972, it made competitors like the Triumph Spitfire and the MG Midget look obsolete, at a stroke.

Bertone styled and built the monocoque structure, and power was by a transverse engine/transmission unit mounted behind the two seats. There was coil spring all-independent suspension, four-wheel disc brakes and good handling to match. It was a credit to the design that customers always seemed to be asking for more power!

The first X1/9s were 1300s, but for 1979 a 1.5 litre engine, allied to a five-speed transmission, was fitted and has been retained ever since. After Fiat pulled out of sports car manufacture in 1982, the X1/9, nominally at least, became a 'Bertone' instead of a 'Fiat'.

The mid-engined X1/9 was given one major up-date when the 1500 engine was introduced, with styling re-touches, in 1978.

FRAZER NASH

Frazer Nash, Bristol-engined family — Produced Isleworth, Middlesex, UK, 1946 to 1956. Production: 95 cars, of all types. **Data for typical model, the Targa Florio** Tubular chassis, with separate light-alloy body shell. Front-mounted ohv six-cylinder Bristol engine, driving rear wheels. 1,971cc, 100bhp/110bhp/120bhp at 5,000/5,500 rpm, to choice. Four-speed gearbox. Transverse leaf and wishbone independent front, torsion bar live axle rear suspension; rack and pinion steering; four-wheel drum brakes; 5.25-16in tyres. Wheelbase 8ft 0in; length 12ft 6in; width 4ft 10in. Unladen weight 1,940 lb. **Derivatives** All cars based on the same chassis design. Engines to various tune, up to 150 bhp for racing. High Speed/Competition/Le Mans Replica type had cycle-type wings and narrow body; Fast Tourer, Targa Florio, Sebring, Mille Miglia types had full-width open or closed coupé styles. **Performance potential** Targa Florio, 100 bhp model: Top speed 114 mph; 0-60mph 10.4 sec; standing ¼-mile 17.8 sec; typical fuel consumption 23 mpg (Imperial).

The post-war Frazer Nash was a completely different car from the 1930s variety, which had in any case petered out by 1939. In a short-lived co-operative effort with Bristol Cars, Frazer Nash evolved a new twin-tube chassis frame and elected to use the Bristol six-cylinder engine (which was, in fact, closely copied from the BMW 328 type).

Frazer Nash output was very restricted between 1948 and 1956, when the Bristol-engined cars were on sale: in total, just 95 cars of all types were made. For one reason, they were extremely expensive, for another they were all very specialized machines, more intended for competition than for normal road use, and in later years the makers, AFN, were increasingly involved with the importing of Porsche cars to the UK.

The Le Mans Replica Frazer Nash was very purposeful, with minimal bodywork.

Frazer Nash owners used their cars on the road and in competition. This is a Mille Miglia model.

The two original models were the full-width Fast Tourer and the stark High Speed, which had separate cycle-type wings and virtually no weather protection. Names, rather than the basic design, altered frequently — the High Speed soon becoming the Competition, and later the Le Mans Replica.

In addition to the Fast Tourer there were later, even more specialized, full-width style cars called Mille Miglia, Targa Florio and finally Sebring models. All had light-alloy coach work, and all were based on the same 8 ft 0in wheelbase chassis, which had independent front but live rear axle suspension. All were bespoke machines, which meant that the customer could order one of several stages of Bristol engine tune. The most 'touring' produced 100 bhp, but up to 120 or even 130 bhp was also available.

There was, incidentally, a single-seater racing car, for use in 2 litre F2 events, which also used the same frame, and had more than 140 bhp — which proves how versatile the Frazer Nash design could be. Although AFN also showed a BMW V8-engined prototype on the same chassis, this never went on sale, and Frazer Nash cars died out as Porsche imports built up.

HEALEY

Produced Warwick, UK, 1946 to 1954. Production: 1,287 of all types including 506 with Nash engines, 756 with Riley engines, 25 with Alvis engines. **Data for Tickford sports saloon** Box-section chassis frame, with separate light-alloy body shell. Front-mounted ohv four-cylinder Riley engine, driving rear wheels. 2,443cc, 104 bhp (gross) at 4,500 rpm. Four-speed gearbox. Coil spring, trailing arm independent front, coil spring live-axle rear suspension; worm-type steering; four-wheel drum brakes. 5.75-15in tyres. Wheelbase 8ft 6in; length 14ft 9in; width 5ft 7in. Unladen weight 2,960 lb. **Derivatives** One basic chassis design throughout, and various bodies, including Westland, Elliot, Silverstone (two-seater), Abbott. Nash-Healey had a 3.8 litre/4.1 litre six-cylinder Nash engine, while Sports Convertible had a 2,993cc/106 bhp (gross) Alvis six-cylinder engine. *No* connection, except in designers, with the later Austin-Healey. **Performance potential** Tickford: Top speed 102 mph; 0-60mph 14.6 sec; standing ¼-mile 19.3 sec; typical fuel consumption 24 mpg (Imperial).

Donald Healey was already a successful rally driver before he entered the motor industry. He became Triumph's technical chief during the 1930s, and after working at Rootes during the Second World War he set up his own small manufacturing business in 1945. The new car he evolved might even have been a Triumph if that company's new owners could have been convinced of its worth — but in the end he settled on using Riley running gear instead.

Every Healey motor car built from 1946 to 1954 used the same basic

The starkly-bodied Healey Silverstone, complete with trailing arm front suspension.

One rare derivative of the Healey was the Alvis 3 litre-engined Sports Convertible.

design of chassis frame, which featured special trailing arm independent front suspension; however, it was always Donald Healey's intention to offer a whole variety of body styles, and this ambition was achieved over the years.

The vast majority of all cars had 2.4 litre Riley engines and gearboxes, and even the capacious saloons could reach 100 mph — which was very creditable indeed for the period in which they were designed and built. The original car was the Westland four-seater tourer, while the two other four-seater dropheads were the Abbott and Sportsmobile models. Two-door sports saloons, with four seats, were the Elliot, Duncan and Tickford.

The most famous Healey was the ultra-sporting Silverstone, which was a stark open two-seater with a very narrow body shell, in which the horizontally-mounted spare wheel also doubled as the rear bumper! Two other Healey derivatives were important. One was the Sports Convertible, which used the same full-width two-seater style as the original Nash Healey, but had an Alvis TA21 3 litre engine and gearbox. The other was the Nash-Healey, which was powered by the six-cylinder American Nash engine and was sold only in the USA.

The Nash-Healey was changed considerably in mid-run, when the wheelbase was lengthened and bodies were fitted to the rolling chassis by Pininfarina in Italy. Not only was the 'Nash' the last of the family to be built, but it was also the most numerous single type. After the Austin-Healey 100 was introduced, the Healey car was obviously on its way out, and assembly at Warwick ended in 1954.

JAGUAR XK SPORTS CARS

Produced Coventry, UK, 1948 to 1961. Production: 12,078 XK120s, 8,884 XK140s, 7,929 XK150s, 1,466 XK150Ss. **Data for XK120** Box-section chassis frame, with steel (early cars alloy) body shell. Front-mounted 2 ohc six-cylinder engine, driving rear wheels. 3,442cc, 160 bhp/180 bhp at 5,000 rpm/5,300 rpm (to choice). Four-speed gearbox. Torsion bar independent front, leaf spring live axle rear suspension; recirculating ball steering; four-wheel drum brakes; 6.00-16in tyres. Wheelbase 8ft 6in; length 14ft 6in; width 5ft 1.5in. Unladen weight 2,910 lb. **Derivatives** Open, fixed-head and drop-head coupé versions of all types. XK140 took over from late 1954, with rack-and-pinion steering, 190 bhp and optional overdrive, XK150 from spring 1957, with new rounded style, optional 210 bhp, optional automatic transmission and four-wheel disc brakes. XK150S, from 1958, had 250 bhp or 3,781cc/265 bhp. All with same basic chassis, and twin-cam XK engine. **Performance potential** XK120 FHC: Top speed 120 mph; 0-60mph 9.9sec; standing ¼-mile 17.3 sec; typical fuel consumption 20 mpg (Imperial). XK150S 3.8 litre: Top speed 136 mph; 0-60mph 7.6 sec; standing ¼-mile 16.0 sec; typical fuel consumption 15 mpg (Imperial).

Originally conceived as a stop-gap project, the Jaguar XK went on to become one of the most glamorous and successful of all British post-war sports cars. The magnificent twin-cam engine was intended for use in a new large Jaguar saloon, but when tooling for this car was delayed William Lyons decided to produce a short-wheelbase version of the new chassis, clothe it in a smart new two-seater body and thus gain valuable

The instantly recognisable shape of the Jaguar XK120 dates from 1948 and hides the famous twin-cam engine.

The XK150 was the third generation of XKs and had fuller, more 'middle-aged' lines than its predecessors.

experience of all the new components — or so he thought. But the sports car enthusiasts would not let him get away with this. The XK120 was shown in October 1948, and even though first deliveries were delayed by nearly a year the demand grew inexorably. At first the body tooling was somewhat rudimentary, but Jaguar were delighted to have to re-think this, with long-term production in mind.

At first the XK120 was sold only as an open two-seater, and most deliveries went to the USA. Its 160 bhp 3.4 litre engine gave it a top speed of more than 120 mph, which was quite sensational by the standards of the day (especially at the price asked), though there was still plenty of scope for improving the steering, the handling and the braking. Before the XK120 was succeeded by the XK140 in 1954, a fixed-head coupé and a drop-head model were both introduced.

The XK140 not only had rack-and-pinion steering, but the coupé and convertible versions also had 2+2 seating. As the years passed engine tune-up kits, optional overdrive and optional wire spoke wheels were all made available and XKs were successful in racing and rallying.

The XK150 of 1957, still on the same basic chassis and running gear, had disc brakes all round and a more roomy though somewhat less sleek body style, but was the best-equipped XK yet. From 1958, too, there was the triple-carb XK150S derivative, and when this was up-graded to 3.8 litres in 1959, with a 136 mph top speed, the Jaguar sports car was a truly formidable performer.

To succeed it, something sensational would be needed — and the E-Type filled the bill perfectly!

Top *When Lancia decided to produce a new rally car, they did it properly! The Stratos had Bertone styling, with a Ferrari Dino engine, and was a huge success story.*

Above *The Lotus Esprit, unveiled in 1975, was a combination of Giugiaro styling with Lotus mid-engined engineering.*

Below *The sleek 1950s-style Lotus Elite had a glass-fibre monocoque, and a Coventry-Climax engine.*

Above *Spot the replica? The genuine MG TF of 1953-55 is in the foreground, the Naylor TF1700 of the mid-1980s behind it.*

Below *MG's timeless MGB style, produced from 1962 to 1980, sold more than half a million examples.*

Above *Shoehorn a Rover V8 engine into the MGB's structure and the result was the MGB GT V8, with a top speed of nearly 130 mph.*

Below *The 1980s style Morgan looks much like its 1930s ancestor — this is the V8-engined Plus 8, with non-standard wheels.*

Bottom *Perhaps the most significant modern US car is the mid-engined Pontiac Fiero. It sells, and sells ...*

Top *Introduced in 1963 and still seliing (with more than a quarter of a million built), the Porsche 911 is a legend. This is the fabulous Turbo model, first seen in 1975.*

Above *The 924 set the front-engined Porsche trend — this is the re-engined, re-touched 944 which followed.*

Below *The Porsche 928S of the late 1970s had 300 bhp, a top speed of at least 150 mph and impeccable road manners. It may be built for 20 years.*

JAGUAR E-TYPE FAMILY

Jaguar E-Type (6-cylinder) family — Produced In Coventry, UK, 1961 to 1971. Production: 15,496 3.8 litre, 41,724 4.2 litre models. **Data for 3.8 litre model** Centre and rear unit-construction pressed-steel body/chassis assembly, with bolt-on multi-tubular front chassis. Front-mounted 2 ohc six-cylinder engine, driving rear wheels. 3,781cc, 265 bhp (gross) at 5,500 rpm. Four-speed gearbox. Torsion bar front independent, coil spring and wishbone independent rear suspension; rack and pinion steering; four-wheel disc brakes; 6.40-15in tyres. Wheelbase 8ft 0in; length 14ft 7.3in; width 5ft 5.25in. Unladen weight 2,625 lb. **Derivatives** Open and fastback/hatchback versions of 3.8 litre. 4.2 litre, with 4,235cc/265 bhp (gross) took over in late 1964, plus longer-wheelbase 2+2 hatchback coupé in mid-1966, with optional automatic transmission (only on this model). **Performance potential** 3.8 litre: Top speed 153 mph; 0-60mph 7.2sec; standing ¼-mile 15.1sec; typical fuel consumption 20 mpg (Imperial). 2+2 version of 4.2 litre: Top speed 139 mph; 0-60mph 7.4sec; standing ¼-mile 15.4sec; typical fuel consumption 21 mpg (Imperial).

Jaguar E-Type (V12-cylinder) family — Produced In Coventry, UK, 1971 to 1975. Production: 15,287 all types. **Data for V12 S3 model** As six-cylinder car except for front-mounted ohc V12 engine. 5,343cc, 272 bhp at 5,750rpm. Optional automatic all types. E70VR-15in tyres. Wheelbase 8ft 9in; length 15ft 4.3in; width 5ft 6.1in. Unladen weight 3,230 lb. **Derivatives** All one wheelbase, but open two-seater or coupé/hatchback 2+2-seater types. **Performance potential** V12 type: Top speed 142 mph; 0-60mph 6.8 sec; standing ¼-mile 14.6sec; typical fuel consumption 16 mpg (Imperial).

According to industrial logic, Jaguar's E-Type was a successor to the D-Type — but how could a road car take over from a sports racing car? The simple explanation is that the E-Type which was put on sale in 1961 was not at all the same car as that conceived in 1956/57!

Even after the headlamp perspex covers had been removed, the E-Type still had a remarkably pure shape.

Jaguar started designing a racing successor to the D-Type just before Sir William Lyons withdrew from 'works' competition. Even at that stage the new car, always called the E-Type, had all-independent suspension and a simplified form of monocoque structure allied to a multi-tubular front end which the D-Type had employed. In the years which followed, the engineers persistently 'productionized' the E-Type, from a racing sports car to a road sports car, and it finally appeared in 1961.

At the time the E-Type was, quite literally, the most newsworthy sports car in the world, combining startling styling with a near-150 mph top speed and a remarkably low price tag. Only Ferrari, at astronomically higher prices, could match its performance. The E-Type not only offered the choice of open or fastback two-seater styles, allied to four-wheel disc brakes and a comfortable ride, but it used the same, relatively low-stressed, XK engine as found in the XK150 sports cars and the new MK 10 saloons.

The only complaints aimed at the E-Type, at first, were that its interior was quite cramped and that its roadholding did not match its looks. Later it also became clear that, if neglected, it could also rot away at an alarming rate.

Nothing, however, could detract from its good looks, or from its colossal performance. When it inherited the larger 4.2 litre engine *and* a new all-synchromesh gearbox for 1965, it became an even better proposition and the launch of a longer-wheelbase 2+2 model in 1966 rounded out a remarkable range.

From this point on the onslaught of USA regulations gradually dulled

Two-seater FHC E-Types could reach well over 140 mph — this is a Series II 4.2 litre.

Top *There were two types of V12 S3 cars, both on the longer wheelbase —the Roadster and the Coupé.*

Above *The S3 V12 E-Type had slightly flared wheel arches and a large air intake to help cool the engine.*

the sharp edge of E-Type performance. Not only did the engine have to be de-tuned, but the nose had to be re-shaped to allow more cooling air into the radiator and the headlamp covers had to be deleted. These, and more, changes were all included in the Series 2 in the autumn of 1968, which might have been environmentally more acceptable in the USA but were by no means as exciting as fast cars.

The final flowering of the E-Type was as the Series III of 1971, when the new light-alloy V12 engine was fitted instead of the old six-cylinder XK and when both styles used the longer wheelbase tub. Fatter tyres and flared wheel arches didn't help the looks, but the acceleration (aided by 272 bhp (DIN)) was better than ever, and the car continued to sell well until 1974/75, when it was dropped, ahead of the launch of the utterly different XJ-S.

JENSEN-HEALEY

Produced At West Bromwich, UK, 1972 to 1976. Production: 10,926, including 473 Jensen GTs. **Data** Unit-construction, pressed-steel body/chassis assembly. Front-mounted 2ohc four-cylinder Lotus engine, driving rear wheels. 1,973cc, 140 bhp at 6,500 rpm. Four-speed box with overdrive (later cars five-speed gearbox). Coil spring independent front, coil spring live axle rear suspension; rack and pinion steering; front disc brakes, rear drums; 185-13in tyres. Wheelbase 7ft 8in; overall length 13ft 10in; width 5ft 3.2in. Unladen weight 2,340 lb. **Derivatives** Jensen GT, announced 1975, had 'sporting estate' style roof on same basic Jensen-Healey monocoque. **Performance potential** Jensen-Healey: Top speed 119 mph; 0-60mph 7.8 sec; standing ¼-mile 16.2sec; typical fuel consumption 24 mpg (Imperial).

The birth of the Jensen-Healey was a direct result of British Leyland killing off the Austin-Healey marque. Healey got together with Kjell Qvale from the USA, who took over Jensen of West Bromwich and with it the capacity to build a sports car.

After dabbling with Vauxhall running gear throughout, the company settled on the brand-new Lotus 16-valve engine, a Chrysler transmission and Vauxhall suspensions, all hidden under a conventionally styled two-seater sports body. Teething troubles from the Lotus engine did not help, neither did corrosion tendencies, though a change to the Getrag gearbox in 1974 was an improvement. The subsequent Jensen GT 'sporting estate' derivative came too late to boost sales, and the last of these strangely anonymous cars was built in 1976, the 'Healey' name dying with it.

The smooth nose of the Jensen-Healey hid a 16-valve 2 litre Lotus engine giving 140 bhp.

LAMBORGHINI MIURA

Produced At S'Agata, Italy, 1966 to 1973. Production: 763 cars, of all types. **Data for Miura P400S** Pressed-steel platform chassis frame, with steel body welded to it on assembly. Mid, transversely-mounted, 2 ohc V12 engine, driving rear wheels. 3,929cc, 370 bhp at 7,700 rpm. Five-speed gearbox. All-independent suspension by coil springs and wishbones; rack and pinion steering; four-wheel disc brakes; GR70VR-15in tyres. Wheelbase 8ft 2.5 in; length 14ft 3.5in; width 5ft 11in. Unladen weight 2,850 lb. **Derivatives** P400 was the original Miura, with 350 bhp, P400S took over in 1969, then 385 bhp SV model from 1971. **Performance potential** Miura P400S: Top speed 172 mph; 0-60 mph 6.7 sec; standing ¼-mile 14.5sec; typical fuel consumption 14 mpg (Imperial).

There is no doubt that the mid-engined Miura caused a stir when it was first seen as a rolling chassis at the 1965 Turin show. No-one truly believed that it would ever be put on sale —not, that is, until a complete car, with that unmistakable Bertone coupé styling, appeared at the Geneva show a few months later!

The Miura's chassis was not merely astonishing because it was mid-engined, nor even because a four-cam V12 engine was used. The real surprise was that this magnificent 4 litre/350 bhp unit was transversely positioned, with a new and unique five-speed transmission into the bargain.

The complete Miura broke new ground in every way. At a stroke it made so-called Ferrari and Maserati 'Supercars' look old-fashioned. Not only that, but road-testers soon found that the claimed power output was very geniune indeed, that the maximum speed was well over 170 mph

The sensational-looking Miura had a V12 engine behind the seats and a 170 mph-plus top speed.

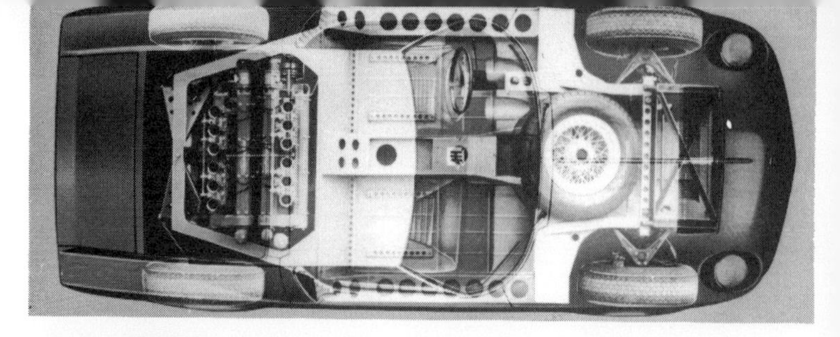

Top *Inventive design packaging saw the Miura's V12 engine mounted across the frame behind the seats.*

Above *The whole of the nose and the complete tail of the Miura could be lifted to attend to servicing needs.*

and that the roadholding from the racing-standard structure and suspension was every bit as good as they hoped. There was no doubt that the Miura was the very best of the fastest Italian cars of the day.

Lamborghini had made no compromises to the design to make it a more civilized road car. There was a lot of noise transmitted to the cabin and there was precious little stowage space. On the other hand, the Miura's looks were quite peerless and the overall character exhilarating. The fact that there was a distinct tendency for the nose to lift at top speed merely added to the excitement, and few owners dared venture up close to that point in any case.

Miuras sold remarkably well, which encouraged Lamborghini to make them even better as the years passed by. To follow the original there was the P400S of 1970, with 370 bhp, while the final P400SV of 1971-73 had no less than 385 bhp and a top speed thought to be around 180 mph.

It was going to take a lot to replace the Miura, but the Countach which superseded it in 1974 was equal to the task.

LAMBORGHINI URRACO

Produced At S'Agata, Italy, 1970 to 1979. Production: 520 P250s, 66 P200s, 190 P300s and 52 Silhouettes. **Data for P250 Urraco** Pressed-steel platform chassis frame, with steel body welded to it on assembly. Mid, transversely-mounted, ohc V8 engine, driving rear wheels. 2,463cc, 220 bhp at 7,500 rpm. Five-speed gearbox. All-independent suspension by coil springs and MacPherson struts; rack and pinion steering; four-wheel disc brakes; 205VR-14in tyres. Wheelbase 8ft 0.5in; length 13ft 1.2in; width 5ft 9.3in. Unladen weight 2,885 lb. **Derivatives** P250 was the original, but from 1974 the P200 (1,994cc/182 bhp) and P300 (2,996cc/2ohc/265 bhp) versions joined the range. The Silhouette of 1976 was Urraco based, but with changed style, only two seats and the 3 litre engine. **Performance potential** Urraco P250: Top speed 143 mph; 0-60mph 8.5 sec; standing ¼-mile 16.6 sec; typical fuel consumption 18 mpg (Imperial). Urraco P300: Top speed 158 mph; 0-60mph 7.6sec; standing ¼-mile 15.6 sec; typical fuel consumption 15 mpg (Imperial).

Strictly speaking, one must view the Urraco as a failure, for it was always intended to outsell cars like the Ferrari Dino and never came close to meeting this target. Even so, many people think it was a more attractive looking machine and that it had more practical accommodation.

The Urraco was Lamborghini's long-rumoured 'small' car. It first appeared in 1970, but staggered into production in 1972. In general layout it was like the Miura, with a transverse mid-mounted engine, but this time there was an all-new single-cam V8 behind the seats, the whole car being smaller but with 2+2 seating and with rather sharper Bertone styling. In the beginning, of course, it was not nearly as fast as other

The Urraco had smooth lines and 'almost' four-seater accommodation.

It was easier to see out of a Urraco than to look into the cockpit. There was a V8 engine behind the seats.

Lamborghinis — though a 143 mph top speed was respectable by almost anyone else's standards!

Like all the best Italian supercars, the Urraco not only had good looks and a fine performance, but an excellent chassis, roadholding and brakes. If only the quality control and after-sales service had been as good as the rest, Lamborghini would surely have had a winner in their stable.

In 1974 Lamborghini also produced the P200 and the P300 derivatives. The P200 was an Italy-only 'taxation special', with 1,994cc and 182 bhp, but the P300 was an ambitiously re-designed car complete with twin-cam (not single cam) cylinder heads, 3 litre capacity and 265 bhp. This made it a near-160 mph machine and a much more formidable proposition.

By this time Lamborghini had gone through several changes of management and were in difficult financial and commercial straits so the Urraco's career suffered. In 1976, however, the Silhouette appeared, effectively a Urraco with a much-modified body style (still by Bertone), only two seats and the 3 litre engine.

The next version of the same basic design, which went on sale in 1981, was the Jalpa — in effect a grown-up Silhouette. The engine displaced 3.5 litres and the car was now much faster than the designers had ever envisaged in 1970.

LAMBORGHINI COUNTACH

Lamborghini Countach family — Produced S'Agata, Italy, 1974 to date. Production: 150 LP400, 385 LP400S, approx 350 LP500S. Quattro valvole production continuing. **Data for LP400** Multi-tubular chassis frame, with separate steel and light-alloy body shell. Mid, in-line, 2 ohc, V12 engine, driving rear wheels. 3,929cc, 375bhp at 8,000 rpm. Five-speed gearbox. All-independent suspension by coil springs and wishbones; rack and pinion steering; four-wheel disc brakes; FR70VR-15in tyres. Wheelbase 8ft 0.5in; length 13ft 7in; width 6ft 6.7in. Unladen weight 3,000 lb. **Derivatives** LP400 was the original Countach, the Countach S following in 1978, with chassis changes, fatter tyres and rear aerofoil. LP500S of 1982 had 4,754cc/375 bhp. **Performance potential** Countach LP400: Top speed 175 mph; 0-60mph 5.6 sec; standing ¼-mile 14.1 sec; typical fuel consumption 11 mpg (Imperial).

Lamborghini Countach quattro valvole — Data As for previous Countachs, except for: 5,167cc, 455bhp at 7,000 rpm. 225VR-15in/345VR15in tyres. Unladen weight 3,188 lb. **Performance potential** quattro valvole: Top speed 178 mph; 0-60mph 4.9sec; standing ¼-mile 13.0sec; typical fuel consumption 15 mpg (Imperial).

It was no wonder that Bertone's own staff gave the new mid-engined Lamborghini its name — for 'Countach' is an expression of amazement in Northern Italy and the new LP500 model was certainly a truly amazing machine. The name stuck, even though the production car was not quite as awe-inspiring as the prototype had been.

Elegant clothes with the elegant Countach, but do the two styles really go together?

The Lamborghini Countach had a unique door-opening arrangement: V12 engine was longitudinally-mounted behind the seats.

The Countach came along, originally in prototype form in 1971, to take over from the Miura. Although the same famous four-cam Lamborghini V12 engine was retained, the rest of the car was different. There was a multi-tubular chassis frame and, although the engine was still mid-mounted behind the seats, it was in the longitudinal position with the gearbox ahead of it, almost between the seats themselves.

The rest of the layout was positively conventional by Lamborghini standards and the Bertone body style was brutally sharp-edged, not nearly as flowing in line as the Miura had been.

At first the LP400 4 litre Countach could reach perhaps 175 mph (though original claims for the 5 litre prototype had spoken of 200 mph!) and this was quite enough for most of the clientele. However, with Ferrari's Boxer coming into the market, and with the onset of exhaust emission laws in various territories, the later production cars actually seemed to be slower than this.

The Countach S — introduced in 1978 — merely had fatter tyres and better aerodynamics, but the LP500S of 1982 had 4.7 litres and its performance restored. The remarkable 'quattro valvole' model of 1985, however, matched the latest Ferrari Testarossa in every way, for with 5.2 litres and 455 bhp it had simply phenomenal performance, stunning looks and unlimited motoring 'sex appeal'. Even so, the laws of aerodynamics could not be cheated and the top speed stayed stubbornly under 180 mph!

LANCIA STRATOS

Produced In Turin, Italy, 1974 and 1975. Production: 492 cars. **Data** Pressed and fabricated steel unit-construction body/chassis assembly. Mid, transversely-mounted 2ohc V6 Ferrari engine, driving rear wheels. 2,418cc, 190 bhp at 7,000 rpm. Five-speed gearbox. Coil spring independent front, coil spring MacPherson strut independent rear suspension; rack and pinion steering; four-wheel disc brakes; 205VR-14in tyres. Wheelbase 7ft 1.8in; length 12ft 2.1in; width 5ft 8.9in. Unladen weight 2,160 lb. **Derivatives** None. **Performance potential** Top speed 143 mph, 0-60mph 6.8 sec. No other authoritative statistics available.

The Stratos was an 'homologation special', pure and simple. Lancia designed the car to be supreme in international rallies, which it was for several years. With much technical help from Abarth and a V6 engine/transmission unit lifted from the Ferrari Dino 246 it had the best possible start.

The car was very short and stubby, but was most attractive, and had very 'nervous' handling habits. Its Bertone-designed and built monocoque was extremely strong, and was clothed with front and rear glassfibre body sections. Accommodation was limited (even finding space for crash helmets was a squeeze) and the car was certainly noisy, but there was plenty of charisma to make up for it all.

Never the world's best road car, the Stratos is now a highly-prized 'collector's piece'. All those rally victories, especially in Monte Carlo, make sure of that.

The mid-Ferrari-engined Stratos enabled Lancia to dominate rallying in the 1970s.

LANCIA MONTE CARLO

Produced In Turin, Italy, 1975 to 1984. Production: 7,595 cars, including USA Scorpion models. **Data for 2 litre model** Unit-construction pressed-steel body/chassis assembly. Mid, transversely-mounted 2ohc, four-cylinder engine, driving rear wheels. 1,995cc, 120 bhp at 6,000 rpm. Five-speed gearbox. Coil spring and MacPherson strut independent suspension front and rear; rack and pinion steering; four-wheel disc brakes; 185HR-13in tyres. Wheelbase 7ft 6.5in; length 12ft 6.1in; width 5ft 5.8in. Unladen weight 2,290 lb. **Derivatives** 1,756cc version ('de-toxed') sold in the USA, badged as a Lancia Scorpion, with only 84 bhp. **Performance potential** 2 litre European version: Top speed 119 mph; 0-60mph 9.8sec; standing ¼-mile 16.0sec; typical fuel consumption 28 mpg (Imperial).

The Monte Carlo was one of those 'nearly' cars which was almost a great success, but which eventually died of neglect — by Lancia and by its clientele. Designed immediately after the Fiat X1/9 (it might once have been badged as a Fiat, and carried the X1/20 project number), it had the same transverse mid-engine layout, all-independent suspension and well laid out two-seater accommodation.

Styling and body construction was by Pininfarina and incorporated a choice of fixed or fold-back roof styles. There were 'sail' panels behind the doors and a rather bluff nose. In the USA the car was called the Scorpion and had a 'detoxed' 1.8 litre engine.

Perhaps the lack of performance in US-market form was one snag: certainly a lack of detail development was another. The Monte Carlo has died away and has not been replaced.

The Lancia Monte Carlo had its engine in the tail, driving the rear wheels. This was the 1980 version.

LOTUS ELITE (1950s TYPE)

Produced At Cheshunt, UK, 1958 to 1963. Production: 988 cars. **Data** Unit-construction glass-fibre monocoque, with some steel reinforcement. Front-mounted ohc four-cylinder engine, driving rear wheels. 1,216cc, 71 or 83bhp at 6,100rpm/6,250rpm. Four-speed gearbox. Coil spring independent front, coil spring MacPherson/Chapman strut independent rear suspension; rack and pinion steering; four-wheel disc brakes; 4.90-15in tyres. Wheelbase 7ft 4in; length 12ft 4in; width 4ft 11.25in. Unladen weight 1,455 lb. **Derivatives** All cars with same style, but engines up to 100 bhp eventually made available. **Performance potential** 83 bhp model: Top speed 118 mph; 0-60mph 11.0sec; standing ¼-mile 17.5sec; typical fuel consumption 35 mpg (Imperial).

In almost every way the original Lotus Elite was startlingly different. It was Lotus's first true road car and certainly the world's first glass-fibre monocoque design. If only Lotus had been able to make it a bit more civilized it might have become a trend-setter.

The engine was by Coventry-Climax, but the all-independent suspension was a real Lotus/Colin Chapman innovation. The car was very light, aerodynamically efficient and fast, but could also be surprisingly economical. There were, however, problems with ventilation (the door windows could not be retracted), with interior noise and with build quality. It seems, too, that Lotus dared not charge enough to make it sufficiently profitable to them. Nevertheless, the Elite set new roadholding standards and the Elan which followed benefited from all the experience gained.

The Lotus Elite of the 1950s was unique in the use of a glass-fibre monocoque. It had a very wind-cheating shape.

LOTUS ELAN AND PLUS 2

Produced At Cheshunt, UK, and Hethel, UK, 1962 to 1974. Production: 9,659 Elans and 4,798 Elan Plus 2s, of all varieties. **Data for Elan 1600 S1 of 1963** Separate steel backbone chassis frame, with glass-fibre body shell. Front-mounted 2ohc four-cylinder engine, driving rear wheels. 1,588cc, 105 bhp at 5,500 rpm. Four-speed gearbox. Coil spring independent front, coil spring MacPherson/Chapman strut independent rear suspension; rack and pinion steering; four-wheel disc brakes; 5.20-13in tyres. Wheelbase 7ft 0in; length 12ft 1in; width 4ft 8in. Unladen weight 1,515 lb. **Derivatives** Fixed head coupé version from 1965, and longer wheelbase/wider Elan Plus 2 from 1967. SE versions had 115 bhp, and Elan Sprint/Elan Plus 2S 130 had 126 bhp engines. **Performance potential** Elan S1: Top speed 115 mph; 0-60mph 8.7sec; standing ¼-mile 16.4sec; typical fuel consumption 30 mpg (Imperial). Elan Plus 2S 130/5-speed: Top speed 120 mph; 0-60mph 7.5sec; standing ¼-mile 16.0sec; typical fuel consumption 28 mpg (Imperial).

By 1960, Lotus were already reeling under the financial disaster of the original Elite and Colin Chapman resolved to replace it with a similar, cheaper and altogether more practical two-seater. This car, which appeared in 1962, was the Elan, the first of a series of twin-cam-engined machines which put Lotus on the road towards prosperity.

The two features which made the Elan such an important car were the use of a steel backbone chassis frame and the engine itself. The backbone frame was not only simple to produce, but it was an ideal layout to enclose the front engine, the final drive unit and the all-independent suspension and steering systems.

The engine was a twin-overhead-cam design, effectively a conversion of the new overhead-valve Ford Anglia/Classic unit, and in one form or another it was to power most new Lotus models produced in the next decade. Harry Mundy (then *Autocar's* technical editor, but one-

The Elan was built either as an open sports car, or as a smart two-seater coupé. This was the S4 version.

To complement the Elan, Lotus produced the Plus 2, with longer wheelbase, wider track and 2+2 seating.

time chief designer at Coventry Climax), designed the engine which, when properly built and maintained, was rugged and ultra-reliable.

The Elan had a glass-fibre body, a soft ride, but quick steering and leech-like roadholding. At first it was sold only as an open two-seater, but from 1965 a smart hardtop version was added and both types were then sold to the end of its career.

In 1967 the Elan was joined by the aptly-named Elan Plus 2, for this was really the same basic design, on a longer wheelbase and wider track, but with a new and elegant 2+2-seater coupé body style. Somehow it was more of a gentleman's carriage than the little two-seater, and it took the transformation of Lotus's image one stage further. In the meantime, incidentally, Lotus had moved its factory from Cheshunt, north of London, to Hethel, near Norwich.

Over the years, both the Elan and the Plus 2 were upgraded and given better equipment, but there was only one important mechanical change. From 1971 both cars were given the 126 bhp 'Big Valve' engine, while the Plus 2S 130 (as it had become called) was also given the new five-speed gearbox from late 1972. Both cars were dropped in the early 1970s to make way for the new 16-valve 2 litre-engined cars which followed.

LOTUS EUROPA FAMILY

Produced At Hethel, UK, 1967 to 1975. Production: 8,975 Europas, all types (some sources quote 9,230). **Data for Europa S1** Separate steel backbone chassis frame, with glass-fibre body shell moulded into position on assembly. Mid-mounted ohv Renault four-cylinder engine, driving rear wheels. 1,470cc, 78 bhp at 6,000 rpm. Four-speed gearbox. Coil spring independent front and rear suspensions; rack and pinion steering; front disc brakes, rear drums; 155-13in tyres. Wheelbase 7ft 7in; length 13ft 0.5in; width 5ft 4in. Unladen weight 1,350 lb. **Derivatives** S2 Europa of 1968 had a bolt-on body. Europa Twin-Cam of late 1971 had 105 bhp Lotus-Ford Twin-Cam in place of Renault engine, and Europa Special (late 1972) had 126 bhp. **Performance potential** Europa-Renault S1: Top speed 109 mph; 0-60mph 10.7sec; standing ¼-mile 17.3sec; typical fuel consumption 26 mpg (Imperial). Europa Special/126 bhp: Top speed 121mph; 0-60mph 7.7sec; standing ¼-mile 15.7sec; typical fuel consumption 28 mpg (Imperial).

Lotus were not the first to produce mid-engined racing cars *or* road cars, but they were the first to do the job properly and to make such cars not only successful but elegantly designed as well. There had been a few mid-engined road cars before the arrival of the Europa, but none so pretty or so neatly engineered.

The Europa was the result of an interesting collaboration with the Renault company, who supplied specially developed engines and transaxles, in return for which the Europa (called 'Europe' on the continent) was to be an export-only car for the first two years. The Europa, therefore, combined well-known Lotus design and manufacturing methods (which included the use of a steel backbone frame and a glass-fibre body shell) with the reputation of Renault's power trains.

In fact, the original Europa was neither very fast nor very well equipped, for with only 78 bhp the top speed was less than 110 mph, the door glasses could not be retracted (only removed altogether), while the body shell was bonded to the chassis on assembly, which made for difficulties if repairs were needed. Even so, the roadholding was well up to expected Lotus standards and no-one was complaining about the general appeal of the styling.

In the years which followed the Europa was progressively improved. The Series 2 of 1968 had its body bolted rather than bonded to the frame and from 1969 it also went on sale in the UK, which boosted sales significantly.

It was, however, always underpowered, and in 1971 the Renault engine was dropped in favour of the Lotus-Ford twin-cam unit — the car logically enough becoming known as the Europa Twin-Cam. At the same time the rear styling was modified to delete the vision-blocking sail panels. A year later the Twin-Cam was replaced by the Europa Special, which had the 126 bhp 'Big Valve' engine. A five-speed gearbox also became optional, and this was later standardized.

The Europa sold well, right to the end — being the last of the Lotus-Ford-powered cars to be manufactured — and it was finally displaced by the much larger and faster Esprit.

Top *The mid-engined Europa was powered by Renault and was at first very much the 'bargain basement' Lotus.*

Centre *There were two luggage compartments in the Europa — one ahead of the driver and one in the extreme tail.*

Bottom *The final version of the Europa family was the Special, complete with 126 bhp twin-cam engine.*

LOTUS ESPRIT AND TURBO

Produced: At Hethel, UK, 1976 to date. Production: 718 S1s, 1,060 S2s, 88 S2.2s, S3 and Turbo still in production. **Data for Esprit S1:** Separate steel backbone chassis frame, with glass-fibre composite body shell, bolted into position on assembly. Mid-mounted 2ohc four-cylinder engine, driving rear wheels. 1,973cc, 160 bhp at 6,200 rpm. Five-speed gearbox. Coil spring independent front and rear suspensions; rack and pinion steering; four-wheel disc brakes; 205HR-14in tyres. Wheelbase 8ft 0in; length 13ft 9in; width 6ft 1.25in. Unladen weight 2,218 lb. **Derivatives:** S2 took over in 1978, and S2.2 (with 2,174cc engine) in 1980. S3 had different chassis and suspension details. Esprit Turbo, 2,174cc with turbocharging, and 210 bhp, entered production in 1980. **Performance potential:** Esprit S2/160 bhp:Top speed 135mph; 0-60mph 8.0sec; standing ¼-mile 16.0sec; typical fuel consumption 21 mpg (Imperial). Esprit Turbo/210 bhp: Top speed 148 mph; 0-60mph 6.1sec; standing ¼-mile 14.6sec; typical fuel consumption 20 mpg (Imperial).

It all started as a styling study by Giugiaro in 1972, on the basis of the existing Lotus Europa, but the Esprit which eventually went on sale in 1976 was different from the Europa in every way except basic configuration. Like the Europa, the Esprit had a mid-engine position, a steel backbone frame, two seats and a glass-fibre body shell — but the differences started from that point.

The engine was the 160 bhp version of the new 2 litre 16-valve Lotus twin-overhead camshaft design, as used in the current Elites and Eclats, and it was mated to the sturdy five-speed transmission of the Citroen SM sports coupé, though this had to be modified as the SM was a front-wheel-drive car with the transmission in its nose!

The mid-engined Lotus Esprit of 1975 had Giugiaro styling and GP-standard roadholding

In the 1980s the normal Esprit was joined by the Esprit Turbo, which had a 210 bhp turbocharged 2.2 litre engine.

Naturally there was all-independent suspension and four-wheel disc brakes, plus the famed (and expected) Lotus combination of a supple ride but truly tenacious roadholding. Naturally, too, the body style only allowed for two passengers and their luggage, but the wedge-nosed shape was simply superb and still looks magnificent today.

Like other new-model Lotuses being introduced in the mid-1970s, the Esprit was expected to go on selling for a number of years. Although it progressed through various 'Marks' in its first decade of life, its shape was still not changed, although the aerodynamic aids were improved from time to time.

The engine tune itself, in normally-aspirated guise, did not alter much, except that in 1980 the fabulous 210 turbocharged model — the Esprit Turbo — came along, and these two types were built side by side.

The original Esprit became the S2 in 1978, and the S2.2 (with 2,174cc) in 1980, then the S3 in 1981 with a new design of chassis and rear suspension, plus bigger wheels and tyres. The Turbo itself donated those tyres to the S3 and its own unique styling features included slats over the engine compartment and extra side skirts.

Quality control was always a problem, but the Esprit's road behaviour was never in doubt. Even today, not many of the world's Supercars can match its handling qualities.

MASERATI 3500GT 6-CYLINDER FAMILY

Produced In Modena, Italy, 1958 to 1969. Production: 2000 3500GTs, 438 Sebrings and 948 Mistrals. **Data for 3500GTI** Multi-tubular chassis frame, with separate steel body shell. Front-mounted 2ohc six-cylinder engine, driving rear wheels. 3,485cc, 235 bhp at 5,800 rpm. Five-speed gearbox, or three-speed automatic transmission. Coil spring independent front, leaf spring live axle rear suspension; recirculating ball steering; four-wheel disc brakes; 185-16in tyres. Wheelbase 8ft 6.5in; length 15ft 4in; width 5ft 9in. Unladen weight 2,750 lb. **Derivatives** 3500GT was original and 3500GTI had fuel injection. Sebring was Vignale-styled version of 3500GTI. Mistral, bodied by Frua, had coupé and spider versions, and replaced Sebring progressively from 1963. **Performance potential** Sebring model: Top speed 137 mph; 0-60mph 8.4sec; standing ¼-mile 16.0sec; typical fuel consumption 16 mpg (Imperial).

Maserati, like Ferrari, really preferred to go motor racing in the 1950s rather than build road cars, but from 1958 they changed that emphasis and introduced a series of fast front-engined cars with six-cylinder engines. Like Ferrari, they also productionized an existing racing engine for this purpose, rather than produce a new unit for the road cars.

The original model was the 3500GT, which combined a simple multi-tubular chassis frame and beam rear axle with the 3.5 litre twin-cam six which was descended from the recent single-seater and sports racing car designs. In this form the engine produced 220 bhp, so right from the start the Maserati's performance was on a par with that of the comparable Ferrari 250GT.

Maserati did not build their own coachwork and nearly all the closed coupés were by Touring, while from 1959 the open Spiders were by Vignale. From 1962 the 3500GT became the 3500GTI, complete with Lucas fuel injection (the first time this equipment was ever fitted to a production car) and 235 bhp: there was a choice between a ZF manual gearbox or Borg Warner automatic transmission. This was only an interim model, as the next 1962 derivative, still on the same basic chassis, was to be the Sebring coupé, for which both styling and body supply were by Vignale. By this time, too, Maseratis had disc brakes and, although they lacked the 'image' edge on Ferrari, they were just as fast and excitingly engineered. By this time factory production was up to about ten cars every week — probably a match for Ferrari.

No fewer than 2,223 3500GT/3500GTI models were delivered, which makes it the most numerous of all the six-cylinder Maseratis. By the mid-1960s Maserati were working towards producing a range of V8-engined models, but they still had time to produce the Mistral, which used the same basic chassis engineering and mechanical layout as the 3500GTI

Above *The 350GT of 1959 was distinctively Italian, but the Maserati character soon shone through.*

Below *The Maserati 3500GT, seen here at a Geneva show, was sold as a coupé or as a cabriolet.*

Top *The smartest of all the six-cylinder engined Maseratis was the Mistral, with body by Frua.*
Above *The Mistral had a Frua style, but in spite of its size it only carried two passengers.*

and Sebring models but had a choice of coupé or spider coachwork styles by Frua.

The engine of the Mistral was the familiar six-cylinder unit, though even at first it was enlarged to 3.7 litres/245 bhp. Later in the car's run the engine was stretched still further, to 4.0 litres and 255 bhp, but this was the end of the road for a famous design. Body style was distinctive, with a snout-like nose and large rear window: however, it is only fair to all parties to point out that Frua used some common panels, glass and wing lines on the bodies they later supplied to AC for the AC428 model! As far as is known, no-one complained, though the two models were contemporary and selling to the same clientele.

The true merit of this Maserati family was that the engines were all amazingly flexible. It was not necessary to avoid driving any of them in heavy traffic, to keep the revs up, or otherwise to behave like a millionaire 'boy racer'. A number had air-conditioning and all in all they were desirable and practical Supercars.

MASERATI GHIBLI, INDY AND KHAMSIN

Produced Modena, Italy, 1965 to 1982. Production: 1,274 Ghiblis, 1,136 Indy models, approx 450 Khamsins. **Data for 4.7 litre Ghibli** Multi-tubular separate chassis frame, with separate steel body shell. Front-mounted 2ohc V8 engine, driving rear wheels. 4,719cc, 330 bhp at 5,000 rpm. Five-speed gearbox, or three-speed automatic transmission. (Optional 4,930cc engine, 355 bhp/5500 rpm.) Coil spring independent front, leaf spring live axle rear suspension; recirculating ball steering, with power assistance; four-wheel disc brakes; 205VR-15in tyres. Wheelbase 8ft 4.3in; length 15ft 0.7in; width 5ft 10.8in. Unladen weight 2,980 lb. **Derivatives** Ghibli was built to 1973; Vignale-bodied Indy had unit-construction shell, built 1969-75, while Bertone-built Khamsin, also with unit-construction, was made 1972-82. **Performance potential** 4.7-litre Ghibli: Top speed 154 mph; 0-60mph 7.5sec; standing ¼-mile 15.1sec; typical fuel consumption 15 mpg (Imperial).

In the mid-1950s, when Maserati were still actively involved in motor racing, they not only evolved a V12 engine for Grand Prix racing, but also a massive and brutally-powerful V8 for sports car racing. It was a development of this V8 which was so important to their road cars' pedigree in the late 1960s and 1970s.

The first Maserati road car to use the V8 was the limited-production 5000GT (of which only 32 were made) and the next was the rather angular four-door Quattroporte saloon, but the first true Supercar was

The Maserati Ghibli of 1966 had styling by Ghia, massive V8 power up front and quite unmistakable profile.

Top *Though the 1969 Indy looked similar to the Ghibli, it had a different, unit-construction, body structure.*

Above *No need to decorate the elegant lines of the Khamsin with an exotic background — the Bertone lines speak for themselves.*

the startlingly attractive Ghibli of 1966. Like all such Maseratis, the Ghibli had coachwork by a specialist supplier — in this case Ghia, and actually pencilled by Giugiaro.

The Ghibli's pedigree was quite complex. The Quattroporte's tubular, boxed and fabricated chassis came first, the shorter-wheelbase version used under the 2+2 Mexico came next and the even shorter-wheelbase chassis of the Ghibli evolved from that. The Ghibli's style was sensationally sleek and by having an overall length of 15ft 1in for a two-seater it was amazingly spacious.

No-one, however, was complaining about the performance, for Ghiblis were built either with 4.7 litres/330 bhp, or 4.9 litres/355 bhp, and even in 'small-engined' form they could nudge 155 mph. In spite of having beam axle rear suspension, it was still a fine-handling machine which sold remarkably well (1,274 examples) for seven years.

Although it was certainly not a direct replacement for the Ghibli (the two cars were sold side-by-side for several years), the Indy was similar in so many ways. Styling and construction was by Vignale, but the shape was very similar indeed to the Ghibli and there was the familiar running gear — of 4.2 litre or 4.7 litre V8 engines, ZF manual or Borg Warner automatic transmissions and a live rear axle — hidden under the skin.

The important differences were that the Indy had 2+2 seating in a longer body, while for the very first time unit construction of chassis and body were used on a Maserati. This made it certain not only that there would be no 'one-off' body styles commissioned for the car, but that there could be no open version either: open-topped cars, in any case, were in decline at this time.

As the last Ghibli of all was built in 1973 and the first Khamsin was delivered in 1974, the one could be said to have replaced the other and indeed there were many important differences, though the V8 engine and the choice of transmissions remained. The Khamsin's unit-construction body chassis unit was by Bertone, as was the styling, but there was independent rear suspension for the first time on such a Maserati, along with several items of Citroen hardware and plumbing, for this was the period in which the Italian concern was controlled by Citroen of France.

The Khamsin, probably the fastest of all front-engined Maseratis, remained in production until 1982.

The Maserati Khamsin had sharply-detailed styling, real Supercar performance and space for only two passengers.

MASERATI BORA AND MERAK

Produced In Modena, Italy, 1971 to 1983. Production: 571 Boras, and approx 1,700 Meraks. **Data for first 4.7 litre Bora** Unit-construction steel body/chassis assembly. Mid-mounted 2ohc V8 engine, driving rear wheels. 4,719cc, 310 bhp at 6,000 rpm. Five-speed gearbox. Coil spring independent front and rear suspension; rack and pinion steering; four-wheel disc brakes; 215VR-15in tyres. Wheelbase 8ft 6.2in; length 14ft 2.4in; width 5ft 8.1in. Unladen weight 3,210 lb. **Derivatives** Bora had 4,930cc/330 bhp from 1976. Merak was same structure, but with Citroen 2,965cc/190 bhp V6 engine at first. Also 220 bhp Merak SS from 1975, and 170 bhp Merak 2000 (1,999cc) from 1977. **Performance potential** 4.7 litre Bora: Top speed 162 mph; 0-60mph 6.5sec; standing ¼-mile 14.6sec; typical fuel consumption 12 mpg (Imperial).

Like Ferrari, Maserati were quite happy to build a multiplicity of front-engined cars and it was probably because of the success of the Lamborghini Miura that both then turned to producing mid-engined Supercars. Maserati's first, the V8-engined Bora, was a startling car by any standards.

First shown in 1971, the Bora was styled by Giugiaro of Ital Design and powered by the familiar 4.7 litre four-cam V8 engine, linked at first to a five-speed ZF transmission. Naturally it had an ultra-modern chassis and could beat 160 mph with some ease. Of course, it only had two seats and limited stowage accommodation, but then so did the Miura and the forthcoming Ferrari Boxer, so no-one complained.

It was, however, a very expensive Supercar, so to increase sales Maserati then launched a less exotic version at the end of 1972, which they called the Merak and which benefited from their commercial links with Citroen. The structure, its styling and the suspension were all as for the Bora, but the engine was a newly developed 90-degree 2.7 litre V6. The engine, and its related transmission, were also used in the front-drive/front-engined Citroen SM coupé of the period. Because the engine was shorter than the big V8 which it displaced, Maserati modified the passenger cabin and squeezed in two extra (and, candidly, quite useless) 'occasional' seats.

Even though Maserati's links with Citroen were dissolved in 1975, after which Alejandro de Tomaso took charge, the Merak was to stay in production until 1983, though the Bora was dropped in 1979.

In the meantime, the Bora was up-engined (in its final form it had 4.9 litres and 320 bhp), while the Merak received much attention. From 1975 to 1979 there was the 1,999cc Merak 2000, sold only in Italy, while the more powerful 220 bhp Merak SS was added to the range in 1975. Surprisingly, there were few styling differences between Bora and Merak — which no doubt allowed drivers of the Merak 2000 to masquerade as 4.9 litre Bora owners!

Top *The Bora of 1971 was Maserati's first mid-engined road car — a two-seater with 4.7 litre V8 engine.*

Above *The Merak had a mid-mounted V6 engine: this is the more powerful SS version, built into the early 1980s.*

Below *The Bora's smooth lines needed no embellishment. There were two seats and a very high standard of equipment.*

MERCEDES-BENZ 300SL

Produced At Stuttgart, West Germany, 1954 to 1963. Production: 3,250 cars, both types. **Data for 300SL 'gull-wing'** Multi-tubular space-frame chassis, with separate steel body shell. Front-mounted ohc six-cylinder engine, driving rear wheels. 2,996cc, 240 bhp (gross) at 6,100 rpm. Four-speed gearbox. Coil spring independent front, coil spring and swing axle independent rear suspension; recirculating ball steering; four-wheel drum brakes; 6.70-15in tyres. Wheelbase 7ft 10.5in; length 15ft 0in; width 5ft 10in. Unladen weight 2,750 lb. **Derivatives** 300SL 'gull-wing', with lift-up doors, was first, replaced from 1957 by 300SL Roadster, with folding hood and conventional doors, plus low-pivot swing axle and optional hardtop. **Performance potential** 300SL 'gull-wing' — typical: Top speed 129 mph; 0-60mph 8.8sec; standing ¼-mile 16.1sec; typical fuel consumption 21 mpg (Imperial).

Like the Jaguar E-Type, the Mercedes-Benz 300SL was originally conceived as a sports racing car, and like the E-Type it also had quite outstanding performance for the period during which it was built and produced. The 300SL of 1954 was, quite literally, the world's most advanced production car and was priced accordingly!

For 1952 Daimler-Benz produced a sensational new sports racing car, whose chassis frame was a positive cat's-cradle of small-diameter tubes. This was known as a 'space frame', and was both light and rigid. In the 300SL it was also linked to an overhead camshaft six-cylinder

The Mercedes-Benz 300SL had lift-up 'gull-wing' doors and a fuel-injected engine, both futuristic when first seen in 1954.

In 1957, the 300SL Roadster, with soft-top and conventional doors, took over from the 'gull-wing' model.

engine laid well over to reduce bonnet height and there was all-independent suspension.

The most obvious innovation, however, was the use of 'gull-wing' doors, which opened upwards rather than forwards or backwards. This was done because of the demands made on space by the chassis construction, but became an immediate talking point among wealthy enthusiasts and in the media.

The original racing version's engine used carburettors, but the 1954 production car was the first in the world to use direct fuel injection. Once it went on sale, the 300SL was immediately seen to be very fast, though factory claims of 'up to 165 mph' top speeds were never confirmed by independent road testers. It also demonstrated very precarious oversteering roadholding habits if pressed beyond the limits.

In spite of the fragile looks of its multi-tube frame, the 300SL soon proved itself in racing and rallying as a rugged and reliable performer. The high-pivot swing-axle rear suspension, however, was also likely to 'bite back' at inexperienced drivers and was changed when the model was updated in 1957.

At this juncture, the 'gull-wing' body was dropped in favour of a more conventional Roadster style (for which an optional hardtop was also available), there were changes to the layout of the frame tubes to allow front-hinged doors to be fitted and low-pivot swing axle suspension was adopted. In this form the 300SL was built for another six years.

MG T-SERIES FAMILY

Produced Abingdon, UK, 1936 to 1955. Production: 3,003 TAs, 379 TBs, 10,000 TCs, 29,664 TDs and 9,600 TFs. **Data for TC** Separate ladder-style steel chassis frame, with steel body panels on wooden body frame. Front-mounted ohv four-cylinder engine, driving rear wheels. 1,250cc, 54 bhp at 5,200 rpm. Four-speed gearbox. Leaf spring beam front, leaf spring live axle rear suspension; cam-gear steering; four-wheel drum brakes; 4.50-19in tyres. Wheelbase 7ft 10in; length 11ft 7.5in; width 4ft 8in. Unladen weight 1,735 lb. **Derivatives** TA, with 1,292cc/50 bhp, was original, TB had 1,250cc/54bhp engine, TC was little-modified TB. TD had new chassis and coil spring independent front suspension, TF had face-lifted body shell and TF1500 had 1,466cc/63 bhp. **Performance potential** TC model: Top speed 75 mph; 0-60mph 22.7 sec; standing ¼-mile approx 22 sec; typical fuel consumption 31 mpg (Imperial).

In the beginning, the MG Car Co was personally owned by Sir William Morris (later Lord Nuffield), but in 1935 various corporate changes led to the company being absorbed into the Nuffield organization. At the same time the MG design office was closed and a new MG sports car, the TA, was designed at the Morris Motors factory and put on sale in 1936.

The T-Series family spanned 19 years, interrupted by the Second World War, and although there were five types — TA, TB, TC, TD and TF — these were covered by only two chassis and two engines. All had bodies made by the traditional method of fixing steel panels to a wooden skeleton, all were classically styled two-seaters and all had a great deal of that elusive character which is now known as the 'Abingdon Touch'.

The TA of 1936 used a long-stroke engine of old design and at first had a non-synchromesh gearbox; it also had a flexible channel-section frame and leaf spring suspension at front and rear. In 1939 it became the identical-looking TB, with new short-stroke engine and synchromesh gears. A small number of TA and TB drop-head coupé bodies were also built by Tickford for the TA and TB models.

After the war the TB was slightly up-dated, with a wider cockpit and minor suspension changes, to become the TC. This was the first MG model ever to reach the 10,000 production mark and was also the first to be exported in large quantities.

For 1950 there was a new model, the TD, which had a new box-section chassis, independent front suspension, plus rack-and-pinion steering, though the styling — complete with free-standing headlamps and flowing front wings — was much as before. For the first time the T-Series was available with left-hand or right-hand steering.

The TD series was a huge success and to keep it going for two more years it became the TF, with face-lifted front and rear sections. In addition, from mid-1954 the engine was enlarged from 1,250cc to 1,466cc, and peak power rose from 57 bhp to 63 bhp. The TF then gave way to the very different MGA in mid-1955.

Above *The TC (looking so nearly like the TA and TB) sold well between 1945 and 1949, in spite of antiquated styling.*

Below *The TD was the first T-Series to have independent front suspension. It could also be supplied with left-hand drive.*

MG MGA FAMILY

Produced Abingdon, UK, 1955 to 1962. Production: 58,750 MGA 1500s, 31,501 MGA 1600s, 8,719 MGA 1600 MKIIs and 2,111 MGA Twin-Cams. **Data for MGA 1500** Separate box-section steel chassis frame, with steel body shell. Front-mounted ohv four-cylinder engine, driving rear wheels. 1,489cc, 72 bhp at 5,500 rpm. Four-speed gearbox. Coil spring independent front, leaf spring live axle rear suspension; rack and pinion steering; four-wheel drum brakes; 5.50-15in tyres. Wheelbase 7ft 10in; length 13ft 0in; width 4ft 9.25in. Unladen weight 1,988 lb. **Derivatives** MGA 1500 available as open or closed bubble-top coupé. MGA 1600 (1,588cc/80 bhp, with front disc brakes) from mid-1959, MGA 1600 Mk II (1,622cc/86 bhp) from spring 1961. Also MGA Twin-Cam (2ohc, 1,588cc/108 bhp, 4-wheel disc brakes) 1958-60, and De Luxe 1600/1600 Mk IIs with Twin-Cam chassis and pushrod engines **Performance potential** MGA 1500: Top speed 100 mph; 0-60mph 15.0sec; standing ¼-mile 19.3 sec; typical fuel consumption 30 mpg (Imperial). MGA Twin-Cam: Top speed 113 mph; 0-60mph 9.1 sec; standing ¼-mile 18.1 sec; typical fuel consumption 24 mpg (Imperial).

The T-Series had been running successfully for many years when MG designers tried to replace it in 1952. The car they designed, code-named EX175, was refused project approval for two years and eventually appeared in 1955 as the MGA. In the meantime the engine and transmission had been changed, for the T-Series units had been rendered obsolete by the BMC merger which swallowed up Nuffield, Morris and MG, and the BMC B-Series engine was specified instead.

The MGA was distinguished by an extremely rigid chassis frame and a shapely new body style, though it retained the coil spring front

The MGA looked sleek and modern in 1955, when it replaced the craggy old TF. It had a BMC B-Series power train.

The bubble-top MGA Coupé had wind-up windows, curved front and rear screens and was a real gentleman's sports car.

suspension and steering of the old TF. Aerodynamically it was much superior to the old model, and as a consequence it was the first MG road car to have a top speed of 100 mph.

The MGA made its public bow as a racing car at Le Mans, where it performed with honour, and it was such an attractive looking package that it began to sell much better than any previous MG. A year after launch the open Roadster was joined by the svelte fixed-head Coupé, which had wind-up windows and a wrap-around screen and looked extremely elegant.

The MGA stayed in production for seven years and was up-dated several times. The MGA 1600 of 1959 had an enlarged engine and front disc brakes, while the MGA 1600 Mk II of 1961 had an even bigger and more powerful engine. All these cars had splendid roadholding, were very comfortable, safe and strong and had a shape which everyone seemed to like.

From 1958 to 1960 there was also the much more complex MGA Twin-Cam, which used a special twin overhead camshaft engine and had four-wheel Dunlop disc brakes with centre-lock disc wheels. When properly 'on song' the Twin-Cam was a very fast car but there were engine problems in service and it was soon withdrawn. The death of the Twin-Cam indirectly gave rise to the 'De Luxe' models, which had Twin-Cam chassis fittings, brakes and wheels but the conventional ohv engines.

MG MGB FAMILY

Produced Abingdon, UK, 1962 to 1980. Production: 115,898 Mk I Tourer/21,835 Mk I GTs, 271,777 Mk II Tourer/103,762 Mk II GTs and 2,591 MGB GT V8 Coupés. **Data for MGB** Unit-construction pressed-steel body/chassis assembly. Front-mounted ohv four-cylinder engine, driving rear wheels. 1,798cc, 95 bhp at 5,400 rpm. Four-speed gearbox, optional overdrive (standard from '75) or three-speed automatic transmission. Coil spring independent front, leaf spring live axle rear suspension; rack and pinion steering; front disc, rear drum brakes; 5.60-14in tyres. Wheelbase 7ft 7in; length 12ft 9.3in; width 4ft 11.7in. Unladen weight 2,030 lb. **Derivatives** Tourer/Roadster was the original, and fastback/hatchback GT followed in 1965. Mk II of 1967 had all-synchro gearbox and other up-grading. From 1973-76 MGB GT V8 had 3,528cc Rover ohv V8 engine and 132 bhp in MGB GT body shell. **Performance potential** MGB: Top speed 103 mph; 0-60mph 12.2 sec; standing ¼-mile 18.7 sec; typical fuel consumption 25 mpg (Imperial). MGB GT V8: Top speed 124 mph; 0-60mph 8.6 sec; standing ¼-mile 16.4 sec; typical fuel consumption 25 mpg (Imperial).

Although the MGB was an outdated and rather disappointing sports car by the time it was finally dropped in 1980, this was because it had been sadly neglected by the British Leyland management in the 1970s in favour of the Triumph TR7. Even so, it continued to sell well to the very end, breaking every record for the production of an MG sports model.

The MGB, logically enough, was introduced in 1962 to take over from the MGA and used many components — engine, transmission and front suspension — from that car. Its big advance compared with the MGA

The MGB made its bow in 1962, the second monocoque MG sports car. it remained in production for 18 years.

The MGB GT, first seen in 1965, was one of the most elegant MGs ever built, for it had vastly more luggage space than before and was quiet and practical, too.

was that it had an extremely rigid monocoque body/chassis assembly, even better roadholding than before, plus a pleasing style and excellent equipment which included wind-up windows for the first time on an MG sports car.

This great car, which the American buyers clearly liked very much, became an even more attractive proposition from 1965 when the GT derivative was put on sale, for that had a permanent roof, a hatchback and even very restricted '+2' seating. It was slightly heavier than the Roadster, but was a very smart and versatile proposition, especially in countries where the sun did not shine all the time.

Right from the start the MGB had a 100+ mph top speed and was competitive, both on performance and price, with rivals like the Triumph TR4. In the next few years it was only necessary to keep building enough cars to satisfy the demand. For 1968, and at the same time as the six-cylinder MGC was put on sale, the Mk II MGB appeared, complete with an all-synchromesh gearbox and with an automatic transmission option, though the latter did not sell strongly.

Thereafter few mechanical, though regular cosmetic, changes were made to the car and in the USA its 1.8 litre engine suffered badly, year after year, as strict exhaust emission regulations were enforced. For 1975, too, the styling was badly hit when large black plastic bumpers were fitted and the whole car raised — once again to meet new rules. It was all downhill, in public esteem, for the MGB after that.

MG MGC

Produced Abingdon, UK, 1973 to 1976. Production: 2,591 cars. **Data** Unit-construction pressed-steel body/chassis assembly. Front-mounted ohv six-cylinder engine, driving rear wheels. 2,912cc, 145 bhp at 5,250 rpm. Four-speed gearbox, with optional overdrive, or three-speed automatic transmission. Torsion bar independent front, leaf spring live axle independent rear suspension; rack and pinion steering; front disc, rear drum brakes; 165-15in tyres. Wheelbase 7ft 7in; length 12ft 9.2in; width 5ft 0in. Unladen weight 2,460 lb. **Derivatives** Open and GT styles, as for MGB model, on which the basic design was based. **Performance potential** MGC with o/d: Top speed 120 mph; 0-60mph 10.0sec; standing ¼-mile 17.7 sec; typical fuel consumption 19 mpg (Imperial).

By the mid-1960s BMC's Austin-Healey 3000 was coming to the end of its life and to take over from it a new six-cylinder version of the MGB was developed. It was a difficult conversion, for the six-cylinder engine (not the same as that of the Big Healey) was a tight squeeze up front and new torsion bar front suspension was chosen to make the transplant easier.

Compared with the MGB, there were only larger wheels and bonnet bulges to give the game away. The weight distribution was not as favourable, but there was a sturdy new all-synchromesh gearbox and an automatic transmission option which was quite popular.

MGCs with overdrive were best of all, very fast and long-legged in GT form, but compared with the 3000 there was a lack of zest and character and the model was dropped after only two seasons. The MGB GT V8 of 1972 was an entirely different type of car.

To spot an MGC you needed to note the bonnet bulges and the larger 15inch wheels.

MORGAN FAMILY: PLUS 4 TO PLUS 8

Produced Malvern Link, UK, 1950 to date. Production: 3,763 Plus 4s (of which 26 were Plus 4 Plus coupés), approx 5,800 4/4s by mid 1980s (production continuing), and approx 2,700 Plus 8s by mid 1980s (production continuing). **Data for Plus 8** Separate Z-section member chassis frame, with wooden-framed body shell, steel or aluminium skinned. Front-mounted ohv V8 Rover engine driving rear wheels. 3,528cc, 160 bhp at 5,200 rpm (original). Four-speed gearbox. Coil spring/pillar independent front, leaf spring live axle rear suspension; worm and nut steering; front disc, rear drum brakes; 185VR-15in tyres. Wheelbase 8ft 2in; length 12ft 8in; width 4ft 9in. Unladen weight 1,900 lb. **Derivatives** For 1977, Plus 8 given five-speed transmission. Plus 4 was original 'post-war' Morgan, with Standard Vanguard — later Triumph TR — engines. Plus 8 took over in 1968. 4/4 1955 on, had same wheelbase/chassis as Plus 4, and a succession of Ford, later optional Fiat, engines. Plus 4 Plus was Plus 4 with glass-fibre FHC body style. **Performance potential** Plus 8, original: Top speed 124 mph; 0-60mph 6.7 sec; standing ¼-mile 15.1 sec; typical fuel consumption 21 mpg (Imperial).

The problem of finding a new style for the latest Morgan really does not exist, for there have been very few changes in the last 50 years. A Morgan lives in its own time capsule and does not need to follow current fashions. There has never been a Morgan with fins, or with 'this year's' wheels, grille or paint job: having established their own basic style in 1935, the company has not found it necessary to change a lot since then.

The style of a post-war Morgan is unmistakable, whatever the model. This is a 2 litre TR4-engined Plus 4 of the 1960s.

The Plus 8, always recognizable by its alloy wheels, has a Rover V8 engine and has been built since 1968.

Yet today's Morgan is vastly different, even from the 1950 variety, in its running gear if not in its looks. It is true that every 'modern' Morgan has used the same type of chassis, with Z-section side members and sliding pillar independent front suspension, and the same body construction, with steel or aluminium panels on a wooden frame. But engines, transmissions and performance have changed completely.

The first four-wheel Morgan was the 4/4 of 1935/36, but the first post-war model was the Standard Vanguard-engined Plus Four of 1950. This car inherited the more powerful Triumph TR engine during the 1950s, along with a slightly more streamlined front and rear style, front disc brakes and wire spoke wheels. It stayed in production until 1968, joined after 1955 by the Ford-engined 4/4 which, up-rated and re-engined from time to time, is still with us in the mid-1980s.

For a brief and unsuccessful period there was also the Plus Four Plus, which hid the traditional chassis under a glass-fibre full-width body style.

The sensation of 1968, however, was the launch of the Plus 8, to take over from the Plus Four. Instead of the old TR engine there was the Rover 3½ litre V8 unit, which gave phenomenal acceleration without adding to the car's weight. Development of this car continues to this day, for the power output has been increased, a five-speed transmission has been adopted, the track has been widened and rack and pinion steering has been made available.

Add to this the offering of Fiat twin-cam engines as 4/4 alternatives and one can see that Morgan development continues apace, though there are still no plans to change the car's looks!

PONTIAC FIERO

Produced Pontiac, USA, 1983 to date. Production: continuing. **Data for original type** Unit-construction pressed-steel body shell, with advanced composite skin panels. Mid, transversely-mounted, ohv four-cylinder engine, driving rear wheels. 2,471cc, 93 bhp at 4,000 rpm; four-speed gearbox, or three-speed automatic transmission. Coil spring independent front and rear suspensions; rack and pinion steering; four-wheel disc brakes; 185R-13in tyres. Wheelbase 7ft 9.3in; length 13ft 4.2in; width 5ft 8.9in. Unladen weight 2,460 lb. **Derivatives** From late 1984, also sold with V6 engine, 2,838cc/142 bhp. **Performance potential** 4-cyl: Top speed 103 mph; 0-60 mph 10.9 sec; standing ¼-mile 17.8 sec; typical fuel consumption 26 mpg (Imperial). V6 version: Top speed 126 mph; 0-60 mph 8.0 sec; standing ¼-mile 16.2 sec; typical fuel consumption 25 mpg (Imperial).

A mid-engined sports car from a General Motors company — whatever next? Yet it is true — in spite of Pontiac's talk of producing a compact little 'commuter' car, the Fiero is a sports car in the Fiat X1/9 or Lancia Monte Carlo mould, without compromises.

At first the only available engine was Pontiac's venerable 'Iron Duke' four-cylinder unit, with 93 bhp and distinctly leisurely performance, but for 1985 models there was also the option of a 142 bhp V6, when performance was much more exciting. Not only did the Fiero have a very sleek style, but the body skin panels were in advanced composite material and all the expected Detroit-options such as automatic transmission and air-conditioning were also available.

The Fiero had all-independent suspension and well-balanced handling, all of which brought it very close indeed to modern European sports car design standards.

The Pontiac Fiero of 1983 was the first-ever North American sports car with mid-mounted engine.

PORSCHE 356 FAMILY

Produced Stuttgart, West Germany, 1949 to 1965. Production: 76,305 cars, of all varieties. **Data for type 356B 75 bhp model** Unit-construction pressed-steel body/chassis assembly. Rear-mounted ohv flat-four cylinder engine, driving rear wheels. 1,584cc, 75 bhp at 5,000 rpm; four-speed gearbox. Torsion bar independent front, torsion bar and swing axle independent rear suspension; worm-and-peg steering; four-wheel drum brakes; 5.60-15in tyres. Wheelbase 6ft 10.75in; length 13ft 2in; width 5ft 5.5in. Unladen weight 2,030 lb. **Derivatives** Three generations of 356 — original, 356A from 1955, 356B from 1959. Coupés, cabriolets and open Speedsters (no 356B Speedsters). Power outputs from 40 bhp/1,086cc to 115bhp/1587cc 2 ohc, and 130bhp/1,966cc 2 ohc (only on Carrera 2 356B model). **Performance potential** 60 bhp 356A: Top speed 102 mph; 0-60mph 14.1 sec; standing ¼-mile 19.1 sec; typical fuel consumption 34 mpg (Imperial). 90 bhp 356B: Top speed 111 mph; 0-60mph 11.5sec; standing ¼-mile 18.3sec; typical fuel consumption 27 mpg (Imperial).

Well before the first Porsche car was ever built the company's founder, Dr Ferdinand Porsche, had enjoyed a glittering career as a designer. Cars as diverse as the VW Beetle and the Mercedes-Benz SSK sports car were listed to his credit. The Porsche sports car was conceived in Austria after the Second World War, though series production was always concentrated at Stuttgart, in West Germany.

The basis of the first design was the underpan and all the running gear of the rear-engined, air-cooled VW Beetle, on to which a series of aerodynamically efficient body styles were drafted. However, as the

The air-cooled rear-engined Porsche 356 was offered in several guises, including this Cabriolet style.

By the early 1960s the Type 356B was a smart and very well-equipped sports car. This, the 1600 version, had up to 90 bhp and a top speed of more than 110 mph.

years passed, more and more special Porsche components were substituted, the cars became faster and more costly and the Porsche legend developed steadily.

Most Porsches were 2+2 seaters, except that the Speedster was a pure two-seater. The majority were closed fastback coupés, but there were cabriolets, hardtop models and the Speedster itself, which was an open sports car. Type 356 became 356A in 1955, distinguished by its one-piece curved windscreen and greater power, while the Type 356B of 1959 had a more consciously 'styled' nose and yet more power.

The most exciting of all these cars used a brand-new four-camshaft air-cooled Carrera engine, originally a 1.5 litre, but 1.6 litres from 1958 and a full 2.0 litres for a few late-model 356Bs. This was originally a pure racing engine, being complex, noisy and expensive, but it made a fast Porsche even faster for full-blooded road use.

The Type 356 Porsche had an instantly recognizable style which no other car approached or even imitated, and a rather clattery air-cooled engine noise to add to its character. The handling, at first, was 'oversteery' and not at all reassuring, due to the preponderance of weight in the tail, but skilled drivers learned to get the best out of the cars, of which later examples were distinctly more manageable. The Porsche company believed in the layout completely and perpetuated it in the 911 models which took over in the mid-1960s.

PORSCHE 911 FAMILY

Porsche 911 family — Produced Stuttgart, West Germany, 1964 to date. Production: exceeding 250,000, all types, by 1986, and continuing. **Data for first 911** Unit-construction pressed-steel body/chassis assembly. Rear-mounted ohc flat-six cylinder engine, driving rear wheels. 1,991cc, 130 bhp at 6,100 rpm; four-speed or five-speed gearbox; or four-speed Sportomatic transmission. Torsion bar independent front, transverse torsion bar independent rear suspension; rack and pinion steering; four-wheel disc brakes; 165-15in tyres. Wheelbase 7ft 3in; length 13ft 8in; width 5ft 3.5in. Unladen weight 2,200 lb. **Derivatives** After the original came the 2,195cc models in 1969, 2,341cc models in 1971, 2,687cc models in 1973, 2,993cc models in 1975, and 3,164cc models in 1983. Coupé, Targa (opening roof) and full Cabriolet body styling variations, and power outputs ranging from 110 bhp to 231 bhp. Plus turbocharged models (see separate spec). **Performance potential** 911S/160 bhp, 1966 model: Top speed 137 mph; 0-60mph 8.0sec; standing ¼-mile 15.8sec; typical fuel consumption 18 mpg (Imperial). 911 Carrera/3,164cc/1983: Top speed 150 mph; 0-60mph 5.4sec; standing ¼-mile 14.0sec; typical fuel consumption 24 mpg (Imperial).

Porsche Turbo family — Produced Stuttgart, West Germany, 1975 to date. Production: 2,873 3 litre Turbos, approx 9,000 3.3 litre Turbos by mid-1986. **Data for 3 litre Turbo** Unit-construction pressed-steel body/chassis assembly. Rear-mounted ohc flat-six cylinder engine with turbocharging, driving rear wheels. 2,993cc, 260 bhp at 5,500 rpm; four-speed gearbox. Torsion bar independent front, transverse torsion bar independent rear suspension; rack and pinion steering; four-wheel disc brakes; 205 or 225VR-15in tyres. Wheelbase 7ft 5in; length 14ft 2in;width 6ft 0in. Unladen weight 2,703 lb. **Derivative** 3,299cc/300 bhp Turbo took over in autumn 1977. 330 bhp tuned-up version also available for '86. **Performance potential** 3 litre Turbo, 1976: Top speed 153 mph; 0-60mph 6.1sec; standing ¼-mile 14.7sec; typical fuel consumption 20 mpg (Imperial).

Whereas the original Porsche 356 always owed some of its heritage to the VW Beetle, the cars we now know as 911s were all-new in every detail. Although they followed the same basic design layout — monocoque construction, all-independent suspension and an air-cooled engine mounted in the tail — they were completely different in every respect.

When first shown the car carried the name 901, but this had to be changed for trade marking reasons and the prototype of 1963 actually went on sale a year later as the 911. The first car's performance was creditable enough, though by later standards an engine size of 2 litres and 130 bhp looked positively pedestrian.

As with the earlier car, the shape of the 911 was distinctive and never likely to be ignored, while the chassis was meticulously engineered. the engine was a single overhead cam air-cooled flat six, linked (depending on the market) to a four-speed or five-speed all-synchromesh transmission. Independent suspension was by torsion bars all round, the rear in this model having semi-trailing link geometry which was a lot more predictable than that of the old-type 356s. Compared with the

Above *In the mid-1960s Porsche built their now-famous car in 4-cylinder 912 and 6-cylinder 911 form. The body shell gradually became wider and wider over the years.*

Below *By the late 1970s the 911 had grown up to become the 3 litre 911SC, complete with modestly flared wheel arches.*

356s, the interior was much more spacious, with generous 2+2 seating. At one time, indeed, Porsche managed to persuade the sporting authorities that it was a genuine four-seater 'touring car'! At first all cars were closed coupés.

To bridge the gap with the 356 there was also a short-lived interim car called the 912, which was effectively the new 911 fitted with the 1.6 litre 90 bhp flat-four of the old car. The same type of car reappeared for a short while in the late 1970s as an 'exhaust emission' special for sale in the USA, this time with a 2 litre engine size.

Once the 911 had got itself established, and in spite of having a noisy cabin due to the constant drone of the air-cooled engine, it sold just as

Below *The Targa version of the 911 had a roof panel which could be removed completely. Chassis and running gear were the same as those of the Coupé.*

Bottom *The mid-1970s 911 look included a front chin spoiler and an impressive tail spoiler which was standard on the Turbo.*

The 2.7 litre Carrera RS was the first to have an engine lid spoiler, commonly called the 'duck-tail' lid.

fast as Porsche could build the cars and deliver them to the showrooms. It has continued to do this so successfully that the quarter-millionth example was built in 1986, and continuity is promised into the 1990s.

In all that time Porsche carried out continuous development programmes, whether to the bodywork, the detail styling, the engine, the transmission, or to the optional equipment. Life at Stuttgart was never dull and the same car rarely survived for more than two or three years without a major improvement being made.

The original coupé was joined by what became known as the 'Targa' model in 1967, which featured a lift-off roof panel and removable rear window, allied to a permanent roll hoop, but it was not until the early 1980s that a full cabriolet version was also put on sale.

Engines have undergone continuous change, enlargement and boosting of output. The 130 bhp 2 litre was joined by the more highly-tuned 160 bhp 911S engine in 1966, and by the less-specialized 110 bhp 911T unit in 1967. Two years later the 2.2 litre engine size took over and for 1974 it was further enlarged to 2.4 litres.

At the same time the first of the ultra-sporting 2.7 litre 911 Carreras went on sale. Two years after that the Carreras became 3 litre cars when all other 911s went up to 2.7 litres, but it was not until mid-1977 that the humblest Porsche also became a 3 litre. In the meantime, the fabulous

For 1986 the 911 Turbo could be ordered with special nose styling and Sport Equipment fittings.

Turbo (more properly known as the Type 930 Turbo) had been launched and went on sale in 1975. The 3 litre engine was allied to turbocharging, and was at first rated at 260 bhp, but after only three years it was up-gunned to 3.3 litres, giving 300 bhp, resulting in the most accelerative of all rear-engined Porsches so far built.

One more major change to the 'bread-and-butter' 911s was then made, for the 1984 model year, when the engine was boosted to 3.16 litres and 231 bhp — this being more powerful than any previous non-turbocharged derivative — and, thus equipped, the 911 looked well-set for the rest of the 1980s.

During all this time there had also been semi-automatic transmission options, the availability of air-conditioning, wider and ever wider tyres and wheels and regular attention to the styling, not only to cover the wider tyres, but (by adding spoilers front and rear) to improve the aerodynamics, too.

Now, as then, ownership of a 911 is macho-man's highest ambition.

PORSCHE 928 SERIES

Produced Stuttgart, West Germany, 1977 to date. Production: 17,710 928 model; 928S production continuing. **Data for first 928** Unit-construction pressed-steel body/chassis assembly. Front-mounted ohc V8 cylinder engine, driving rear wheels. 4,474cc, 240 bhp at 5,500 rpm; five-speed gear box or three-speed automatic transmission. Coil spring independent front, coil spring and semi-trailing linkage independent rear suspension; rack and pinion steering with power assistance; four-wheel disc brakes; 225/VR-16in tyres. Wheelbase 8ft 2.4in; length 14ft 7.1in; width 6ft 0.3in. Unladen weight 3,200lb. **Derivatives** 928S, announced in 1979, had 4,664cc/300 bhp, and 928S2 of 1983 had 310 bhp. **Performance potential** 928: Top speed 142 mph; 0-60mph 7.5sec; standing ¼-mile 15.7 sec; typical fuel consumption 18 mpg (Imperial). 928S: Top speed 152 mph; 0-60mph 6.2 sec; standing ¼-mile 14.3sec; typical fuel consumption 19 mpg (Imperial).

Porsche remained faithful to the rear-mounted air-cooled engine philosophy for many years, until two different new models —the 924/944 and the 928 — were developed in the 1970s. Although these cars looked superficially similar, they were utterly different in everything except their general layout. Both types used a front-mounted, water-cooled engine and a rear-mounted transmission in unit with the final drive.

The 928 was a bigger and heavier car than the 911 which it was originally expected to replace and did not use any important parts from that famous car. There was a conventional steel monocoque, with a smoothly rounded style, headlamps which stared upwards from the bonnet when not in use, and a useful hatchback access to the luggage compartment. Compared with the 911, too, it was a much more genuine four-seater.

The innovation was all hidden under the skin. Up front was a light-alloy V8 engine (at first of 4.5 litres and 240 bhp), while there was a brand-new

The 928 had ultra-smooth lines, a capacious boot and still found space for four occupants.

Top *By comparison with the well-loved 911, the 928 had a positively understated style, but remarkably good roadholding and refinement.*

Above *The mid-1980s 928 S2 had 4.7 litres, 310 bhp and a top speed of more than 150 mph.*

five-speed transmission (or optional Daimler-Benz automatic) in the transaxle: the two were linked by a stout torque tube which enclosed the propeller shaft.

Naturally there was all-independent suspension, that at the rear being a very cleverly-detailed layout called the 'Weissach' axle, which provided more stable cornering in extreme conditions — for instance if the driver felt obliged to lift-off in mid-corner.

All in all, the 928 was a very Teutonic, very 'efficient' car, which somehow lacked character at first — especially as its 142 mph top speed made it slower than some of the old-fashioned 911 range. Not only did it show no signs of taking over from the 911, but it did not sell particularly well at first.

Accordingly, Porsche hastened to produce the 928S, which had 300 bhp and much more performance, while by the mid-1980s four-valve cylinder heads had also been produced, at first for sale only in the USA. On Porsche's existing record, we may be sure that the 928 — in one developed form or another — will be on sale well into the 1990s.

PORSCHE 924 AND 944 FAMILY

Porsche 924 family — Produced Neckarsulm, West Germany, 1975 to date. Production (by 1986): approx 140,000 924s and 12,114 924 Turbos. **Data for original 924** Unit-construction pressed-steel body/chassis assembly. Front-mounted ohc four-cylinder engine, driving rear wheels. 1,984cc, 125 bhp at 5,800 rpm; four-speed or five-speed gearbox, or three-speed automatic transmission. Coil spring and MacPherson strut independent front, torsion bar and semi-trailing link independent rear suspension; rack and pinion steering; front discs, rear drum brakes; 185HR-14in tyres. Wheelbase 7ft 10.5in; length 13ft 8.2in; width 5ft 5.2in. Unladen weight 2,380 lb. **Derivatives** 924 Turbo (1978-82) had turbocharging, 170 bhp and four-wheel discs, 924 Carrera was short-lived 'homologation special', while 924S for 1986 has 2,479cc/150 bhp engine and four-wheel discs. Also 944 —see separate spec. **Performance potential** 924: Top speed 126 mph; 0-60mph 9.5sec; standing ¼-mile 17.2sec; typical fuel consumption 23 mpg (Imperial).

Porsche 944 family — Produced Neckarsulm, West Germany, 1981 to date. Production: more than 25,000 944s and 944 Turbos by 1986. **Data for 944** Unit-construction pressed-steel body-chassis assembly. Front-mounted ohc four-cylinder engine, driving rear wheels. 2,479cc, 163 bhp at 5,800 rpm; five-speed gearbox or three-speed automatic transmission. Coil spring and MacPherson strut independent front, torsion bar and semi-trailing link independent rear suspension; rack and pinion steering; four-wheel disc brakes; 185VR-15in tyres. Wheelbase 7ft 10.5in; length 13ft 8.2in; width 5ft 8.3in. Unladen weight 2,600 lb. **Derivatives** 944 Turbo of 1985 had turbocharging and 220bhp/5,800 rpm. **Performance potential** 944: Top speed 137 mph; 0-60mph 7.4sec; standing ¼-mile 15.6sec; typical fuel consumption 29 mpg (Imperial).

It was a combination of many events, involving industrial and personal upheavals, which led to this particular project eventually being called a

This is the Porsche 924, which might well have been an Audi sports car if commercial upheavals had not intervened.

Above *The 924 Turbo had the same 2 litre engine, but an output of 170 bhp and a top speed of around 140 mph. Note the extra air scoops.*

Below *The 944 uses the same basic body structure as the 924, but with flared arches and extra aerodynamic aids.*

Porsche rather than an Audi. Like the 928, the car we now call the 924 had a front engine/rear transmission layout, but it was originally designed by Porsche for VW-Audi and was to be built at the ex-NSU factory at Neckarsulm.

Only after VW's fortunes had taken a tumble, new touring cars had been rushed through to succeed the Beetle and new top management had been appointed did the car revert to a 'Porsche' name, though it retained its mainly-Audi mechanical parentage. In the event, it took over for Porsche where the VW-Porsche 914 left off.

At first the 924 was rather an 'ordinary' car by the Stuttgart company's previous standards, for its water-cooled Audi engine produced a mere 125 bhp, top speed was about 125 mph and the styling was smooth but by no means striking. Like the 928, however, it was blessed with torque tube transmission, had very good and predictable roadholding and was always relatively cheap by previous Porsche standards.

In the next few years two rather more fierce 924s appeared, one being the 170 bhp turbocharged model, built for four years, the other the limited-production 924 Carrera GT, which had a 210 bhp turbocharged unit. Both these cars, however, were really stopgaps until the definitive 944 arrived in 1981.

The 944 was much more a 'pure' Porsche, for it had a rugged 2.5 litre four-cylinder engine, with design similarities to the 928's V8 unit and it also had the flared wheel arches and more 'aggressive' style of the previous 924 Carrera. With its 137 mph top speed, better acceleration (and a considerably higher price tag!) it became instantly desirable.

Both types — 924 and 944 — continued to be built successfully and in large numbers, side by side, for they appealed to different types of customer. At this time, indeed, Porsche had their best-ever range of cars — the 924, 944 and 928 with front engines, and the phenomenal 911 and Turbo rear-engined models backing them up. The 924 and the 944 were both ultra-reliable machines, not quite full four-seaters but practical for use as sports cars or business machines.

The 944 engine was clearly capable of a lot more than the initial 163 bhp and in 1985 the thoroughly re-developed 944 Turbo appeared. Not only did this have a lusty 220 bhp engine, but it was equipped with better front and rear aerodynamic spoilers in step with its 150 mph-plus top speed. And even now the 944 range is still in its infancy — consider how good it might be in the 1990s!

SAAB SONETT II AND III

Produced Arlov, Sweden, 1966 to 1974. Production: 1,868 Sonett IIs/V4s and 8,351 Sonett IIIs. **Data for Sonett II** Unit-construction body shell, with pressed-steel understructure and glass-fibre superstructure moulded into place on assembly. Front-mounted two-stroke three-cylinder engine, driving front wheels. 841cc, 60 bhp at 5,200rpm; four-speed gearbox. Coil spring independent front, coil spring 'dead' beam axle rear suspension; rack and pinion steering; front wheel discs, rear drum brakes; 155-15in tyres. Wheelbase 7ft 1in; length 12ft 4.4in; width 4ft 10in. Unladen weight 1,565lb. **Derivatives** Sonett V4 took over in late 1967, with 1,498cc Ford V4 engine/65 bhp. Sonett III, built 1970-74, had restyled body, 1,699cc/75 bhp. **Performance potential** Sonett III: Top speed 100 mph; 0-60mph 14.4sec; standing ¼-mile 19.0sec; typical fuel consumption 29 mpg (Imperial).

Saab originally built only aeroplanes and sold their first front-drive car in 1950. The Sonett was a low-investment sports coupé project using the saloon's underpan and running gear, with stylish glass-fibre superstructure. As the saloons changed, so did the Sonetts, which explains why three distinct types were sold in eight years.

The Sonett IIs either had 3 cylinder two-stroke or (from 1967) V4 engines, both with front-wheel drive, while the much more stylish Sonett III had 1.5 litre or 1.7 litre V4 engines. They were all fastback two-seaters which sold well in the USA, but were always rare in Europe, and demand was still building up when Saab abandoned the project in the face of new difficult-to-meet USA regulations.

Although Saab went on to produce exciting turbocharged engines, they have so far never again built a sports coupé.

The best-selling Sonett was the more sharply styled III, which had Ford-Germany V4 power.

SUNBEAM ALPINE (1953 TYPE)

Produced Coventry, UK, 1953 to 1955. Production: approximately 3,000 cars. **Data** Separate box-section steel chassis frame, with pressed-steel body shell. Front-mounted ohv four-cylinder engine, driving rear wheels. 2,267cc, 80 bhp (gross) at 4,200 rpm; four-speed gearbox (late models also with overdrive). Coil spring independent front, leaf spring live axle rear suspension; worm and nut steering; four-wheel drum brakes; 5.50-16in tyres. Wheelbase 8ft 1.5in; length 14ft 0.25in; width 5ft 2.25in. Unladen weight 2,965 lb. **Derivatives** None — Alpine had evolved from four-door Sunbeam Talbot 90 saloon on same chassis. **Performance potential** Top speed 95 mph; 0-60mph 18.9sec; standing ¼-mile 21.1 sec; typical fuel consumption 24 mpg (Imperial).

Rootes produced a fine series of Sunbeam-Talbot saloons from 1948 onwards, and in 1953 revealed a smart open two-seater version of the design, which they called the Alpine. This was heavy, rather ponderous, a touch underpowered and was afflicted with a steering column gearchange, but it was nevertheless an intriguing export prospect.

It was, however, too expensive to sell well against cars like the Austin-Healey 100 and the Triumph TR2, so its career lasted for only two years. In that short period, though, it undertook several high-speed endurance runs and succeeded twice in the Alpine rally (after which it was named). Alpines starred in more than one glossy Hollywood film and were notable as the first Sunbeams to discard the name 'Talbot' — which came back in 1979.

The next car to carry the Alpine name was a completely different design in every way.

Racing driver Mike Hawthorn and competition team manager Norman Garrad in a Sunbeam Alpine.

SUNBEAM ALPINE AND TIGER

Sunbeam Alpine (1959 type) — Produced Coventry, UK, 1959 to 1968. Production: 11,904 Series 1s, 19,956 SIIs, 5,863 SIIIs, 12,406 SIVs and 19,122 SVs. **Data for Series I** Unit-construction, pressed-steel body-chassis assembly. Front-mounted ohv four-cylinder engine, driving rear wheels. 1,494cc, 78 bhp at 5,300rpm; four-speed gearbox, optional overdrive. Coil spring independent front, leaf spring live axle rear suspension; recirculating ball steering; front wheel discs, rear drum brakes; 5.60-13in tyres. Wheelbase 7ft 2in; length 12ft 11.25in; width 5ft 0.5in. Unladen weight 2,135lb. **Derivatives** SII of 1960 had 1,592cc/80 bhp, SIII of 1963 had 83 bhp, SIVs had style changes and all-synchro 'box, SVs from 1965 had 1,725cc/93 bhp. **Performance potential** Series I: Top speed 98 mph; 0-60mph 14.0sec; standing ¼-mile 19.8sec; typical fuel consumption 26 mpg (Imperial).

Sunbeam Tiger — Produced West Bromwich, UK, 1964 to 1967. Production: 6,495 Tiger I, and 571 Tiger II. **Data for Tiger I** Basically as Sunbeam Alpine, except for ohv Ford-USA V8 engine. 4,261cc, 164 bhp (gross) at 4,400 rpm. No overdrive. Rack and pinion steering. 5.90-13in tyres. Overall length 13ft 2in. Unladen weight 2,525 lb. **Derivatives** Tiger II of 1967 had 4,727cc/200 bhp (gross). **Performance potential** Tiger I: Top speed 117 mph; 0-60mph 9.5sec;standing ¼-mile 17.0sec; typical fuel consumption 17 mpg (Imperial).

Production of a new generation of Rootes cars began in 1955 with the Rapier, Minx and Husky models. The smart, finned Alpine sports car followed in 1959, with the short-wheelbase Husky underpan, the Rapier's running gear and very smart styling. The engine was a 1.5 litre at

The Sunbeam Alpine of 1959 hid its Rapier running gear under this smart two-seater body style.

The Tiger, as exhibited at the 1965 Earls Court Motor Show, had a Ford V8 engine of 4.2 litres.

first, but eventually grew to 1,725cc, when the Alpine V had a 100 mph top speed. In every way this car was a competitor for models like the MGA/MGB and for the Triumph TR of the day.

Like most Rootes cars of the period, the Alpine was solidly engineered and well-equipped, though the roadholding was perhaps slightly less 'sporty' than that of its rivals. Some Alpines had overdrive, many had the optional hardtop, all had wind-up windows and at one time you could even order automatic transmission.

The Tiger was a rumbustious derivative of the Alpine which was only on sale for three years. Like the AC Cobra, it was really a simple transplant of an American Ford V8 engine into the existing structure. Jensen, of West Bromwich, assembled the Tiger on behalf of Rootes and naturally most sales were to North America.

If the Tiger had any disappointments it was that it looked almost exactly like the Alpine (which did nothing for the status-seekers). Even the rare, 1967-only, 4.7 litre Tiger II was not special enough, or fast enough, to match the Cobra.

TRIUMPH TR2-TR6 FAMILY

Triumph TR2 family — Produced Coventry, UK, 1953 to 1962. Production: 8,628 TR2, 13,377 TR3, 58,236 TR3A, 3,331 TR3B. **Data for TR2** Separate box-section steel chassis frame, with pressed-steel body shell. Front-mounted ohv four-cylinder engine, driving rear wheels. 1,991cc, 90 bhp at 4,800 rpm; four-speed gearbox, optional overdrive. Coil spring independent front, leaf spring live axle rear suspension; cam-and-lever steering; four-wheel drum brakes; 5.50-15in tyres. Wheelbase 7ft 4in; length 12ft 7in; width 4ft 7.5in. Unladen weight 1,848 lb. **Derivatives** TR3 of 1955 had 95, later 100bhp and front discs from late 1956. TR3A had restyled nose, optional 2,138cc engine, available from 1958. TR3B, USA only in 1962, had engine choice and all-synchro 'box. **Performance potential** TR2: Top speed 103 mph; 0-60mph 11.9sec; standing ¼-mile 18.7sec; typical fuel consumption 33 mpg (Imperial).

Triumph TR4, TR5, TR6 family — Produced Coventry, UK, 1961 to 1976. Production: 40,253 TR4s, 28,465 TR4As, 2,947 TR5s, 8,484 TR250s, 94,619 TR6s. **Data for TR6** Separate box-section steel chassis frame, with pressed-steel body shell. Front-mounted ohv six-cylinder engine, driving rear wheels. 2,498cc, 150 bhp at 5,500 rpm; four-speed gearbox, optional overdrive. Coil spring independent front, coil spring and semi-trailing link independent rear; rack and pinion steering; front discs, rear drum brakes; 165-15in tyres. Wheelbase 7ft 4in; length 13ft 3in; width 4ft 10in. Unladen weight 2,473 lb. **Derivatives** TR4 was original, with TR3A chassis, 2,138cc/4-cyl engine. TR4A of 1965 had new all-independent chassis and 104 bhp (live rear axle, USA only). TR5 of 1967 had six-cyl/150 bhp engine (TR250 was USA version, with 104 bhp). TR6 later detuned (from 1973) with 124 bhp. **Performance potential** TR6, original: Top speed 119 mph; 0-60mph 8.2sec; standing ¼-mile 16.3sec; typical fuel consumption 22 mpg (Imperial).

Although Standard bought Triumph in 1944 it took them nearly a decade to establish a winning formula for their new marque. It was their inspired decision to produce a lively sports car at a remarkably low price which did the trick.

After a false start at the prototype stage, the TR2 soon made its mark

By 1955 all the best-specified TR2s had optional hardtops and centre-lock wire wheels, plus overdrive, of course.

Top *The most numerous of the 'classic' TRs was the wide-mouthed TR3A, built from 1957 to 1961.*

Above *The Michelotti-styled TR4 was really old wine in a new bottle, for the TR3A's chassis and running gear had been retained.*

Below *For 1968 Triumph offered the TR5, which had a fuel-injected 6-cylinder 2½ litre engine.*

The TR6, built from 1969 to 1976, had TR5's body re-styled by Karmann but the same running gear. US-market cars had carburettors, not fuel injection.

at home and overseas. Although the chassis design was simple — some would say crude, because of the hard ride and the bump steer caused by limited back axle movement — it was also very robust. With the wet-liner engine proving to be both powerful *and* economical, it was not surprising that it began to sell well.

From 1953, when the first deliveries were made, to 1962 the car progressed through TR3, to TR3A and finally to TR3B without the attractive style being changed. Along the way, the power output increased from 90 bhp to 100 bhp, overdrive became optional (and was fitted to many of the cars built), disc front brakes were specified (from 1956/57, making this the first series-production British car to have discs) and many competition successes were notched up.

In 1961 a completely new body shell was fitted over the familiar chassis, this being the Michelotti-styled TR4. It included such advances as wind-up door windows and face-level ventilation. In 1965 the TR4A took over, this looking almost the same but having a new chassis with all-independent suspension.

The famous old 'wet-liner' four-cylinder engine (as also used in the Standard Vanguard, and the Ferguson tractor!) was dropped in 1967 and was replaced by the newer in-line 2.5 litre 'six' — fuel-injected as the TR5 in Britain and Europe, but fitted with carburettors as the TR250, for USA sale alone. Only 18 months later the body shell was substantially restyled by Karmann and subsequently all cars became known as the TR6, whether fitted with fuel injection or carburettors.

The TR6 was the last and the best-selling of all the separate-chassis TRs and is now remembered as a rather 'hairy-chested' sports car, even though its engine was considerably smoother than the 'four' which it replaced. The TR7 which took over from 1975 was a completely different car, not only in concept but in its design and engineering.

TRIUMPH TR7 AND TR8

Produced Liverpool, then Coventry, then Solihull, UK, 1975 to 1981. Production: 112,368 TR7s and 2,722 TR8s. **Data for TR7** Unit-construction pressed-steel body/chassis assembly. Front-mounted ohc four-cylinder engine, driving rear wheels. 1,998cc, 105 bhp at 5,500 rpm;' four-speed, later five-speed gearbox, or three-speed automatic transmission. Coil spring/MacPherson strut independent front, coil spring live axle rear suspension; rack and pinion steering; front discs, rear drum brakes; 175-13in tyres. Wheelbase 7ft 1in; length 13ft 4.1in; width 5ft 6.2in. Unladen weight 2,205 lb. **Derivatives** Original TR7 was a coupé, with convertible from 1979. TR8 was for USA sale, with Rover V8 engine, 3,528cc/133 bhp. TR7 V8 was competition-only version of TR8. **Performance potential** TR7 4-speed: Top speed 109 mph; 0-60mph 9.1sec; standing ¼-mile 17.0 sec; typical fuel consumption 29 mpg (Imperial).

British Leyland decided to produce one new corporate sports car in the 1970s and the Triumph TR7 was the result. Planned, in theory, to take over from the TR6 *and* the MGB, the new wedge-styled monocoque model was intended to have a multitude of engine choices, but strikes and poor sales saw most of these abandoned.

At first there was only a fixed-head coupé, but the convertible arrived in 1979. Most cars went to the USA, in 2 litre form, and most had five-speed, or automatic, transmissions. The BL 'works' rally team used V8-engined cars from 1978, but the Rover-engined TR8 road car was delayed until 1980 and sold only slowly; it was never marketed in the UK.

In spite of its controversial looks, the TR7 steered, handled and rode better than any previous TR. It was a real missed opportunity for BL — one of many in this period.

The smartest of all TR7s was the drophead, built only in the model's last three years.

TRIUMPH SPITFIRE AND GT6

Triumph Spitfire — Produced Coventry, UK, 1962 to 1980. Production, 45,573 Spitfire 1, 37,409 Spitfire 2, 65,320 Spitfire 3, 70,021 Spitfire Mk IV, 95, 829 Spitfire 1500s. **Data for first Spitfire** Separate box-section steel chassis frame, with pressed-steel body shell. Front-mounted ohv four-cylinder engine, driving rear wheels. 1,147cc, 63 bhp at 5,750 rpm; four-speed gearbox, optional overdrive. Coil spring independent front, transverse leaf/swing axle independent rear suspension; rack and pinion steering; front discs, rear drum brakes; 5.20-13in tyres. Wheelbase 6ft 11in; length 12ft 1in; width 4ft 9in. Unladen weight 1,568 lb. **Derivatives** Mk 2 of 1965 had 67 bhp, Mk 3 of 1967 had 1,296cc/75bhp, Mk IV of 1970 had de-tuned engine, but swing-spring rear end, 1500 of 1974 had 1,493cc. **Performance potential** Mk 1: Top speed 92 mph; 0-60mph 17.3sec; standing ¼-mile 20.9sec; typical fuel consumption 33 mpg (Imperial).

Triumph GT6 — Details as for Spitfire except: **Produced** 15,818 Mk 1, 12,066 Mk 2, 13,042 Mk 3, between 1966 and 1973. **Data** Six-cylinder engine, 1,998cc/95 bhp (Mk 1), 104 bhp (others). Transverse leaf/lower wishbone independent rear on Mks 2 and 3. 155-13in tyres. Unladen weight 1,904 lb. **Derivatives** All GT6 cars were fastback/hatchback coupés. **Performance potential** Mk 1: Top speed 106 mph; 0-60mph 12.0sec; standing ¼-mile 18.5sec; typical fuel consumption 24 mpg (Imperial).

The Spitfire and GT6 sports cars evolved directly from the Herald and Vitesse saloons, respectively, which were four-cylinder and six-cylinder versions of the same separate-chassis frame design, with four-wheel independent suspension. Compared with the touring cars, the sports cars had shorter wheelbases and, of course, just two-seat accommodation.

The Spitfire's body style, like that of the touring cars, was by Michelotti, featured a lift-up bonnet/nose/front wings section and was an open tourer with an optional detachable hardtop. Over the years the engines improved from 1,147cc/63 bhp (net) to 1,493cc/71 bhp (DIN), and top speeds rose from 92 mph to more than 100 mph.

The original, pretty Spitfire was announced in 1962 and was based on the chassis engineering of the Herald saloon.

Top *From 1970 the Spitfire was re-skinned, with smoother body lines, different wheels and more effective rear suspension.*

Above *The GT6 had a six-cylinder engine up front and a fastback/hatchback body style, but was closely based on the Spitfire.*

Compared with its obvious rival, the Sprite/Midget, the Spitfire was a more roomy, better trimmed, softer riding car, but it was afflicted with swing axle rear suspension which affected the roadholding somewhat. Not until the later 'swing spring' system was installed for the Mk IV did it become truly competitive.

The GT6, first introduced in 1966, had a fastback fixed head style and the 2 litre six-cylinder engine. From 1968, and the introduction of the Mk II version, there was different 'lower wishbone' rear suspension which improved the roadholding considerably. Some people even went so far as to dub it a 'mini E-Type' — there was a slight resemblance in looks and in character.

TVR FAMILY

Produced Blackpool, UK, from 1958 to date. Production: approx 1,100 Grantura/Griffith, 1,344 Vixen/Tuscan/2500, 3,040 M-Series/Taimar, Tasmin series continuing. **Data for 3000M** Multi-tubular chassis frame, with separate glass-fibre body shell. Front-mounted ohv Ford V6 engine, driving rear wheels. 2,994cc, 142 bhp at 5,300 rpm; four-speed gearbox. Coil spring independent front and rear suspensions; rack and pinion steering; front discs, rear drum brakes; 165-15in tyres. Wheelbase 7ft 6in; length 12ft 10in; width 5ft 4in. Unladen weight 2,240 lb. **Derivatives** Originals were Granturas, with trailing arm rear suspension, Vixens followed in 1967, then M-Series took over in 1972. All-new Tasmin arrived in 1980. **Performance potential** 3000M: Top speed 121 mph; 0-60mph 7.7sec; standing ¼-mile 16.0sec; typical fuel consumption 24 mpg (Imperial).

In spite of several serious financial and personal upheavals, the same breed of TVR sports cars has been built in Blackpool for three decades. In other words, every one has had a multi-tube chassis frame, all-independent suspension and a glass-fibre body shell.

The first TVRs were Granturas, but Vixens and Tuscans followed in the 1960s. The M-Series cars, mainly with Ford or Triumph engines, were built during the 1970s (including a few extremely fast turbocharged Ford-engined models), while the sharply-styled Tasmins ushered in the 1980s. Their descendants are still with us.

All but a very few TVRs have been two-seaters. The first convertibles were built in the late 1970s and these are now the most popular, especially in export markets.

The Vixen of the late 1960s was so typical of all TVR's cars of the period, with glass-fibre body shell and two seats.

GLOSSARY

You've read it all, but is it clear? This is what classic car motoring is all about:

All-independent suspension Independent front suspension to front and rear wheels.
Backbone chassis Structural frame members concentrated near the centre line, or backbone, of the car.
Body/chassis unit Term used to indicate that the two are permanently fixed together on assembly — see 'Monocoque' and 'Unit construction'.
Cabriolet Alternative name for a drop-head coupé, or convertible.
Camber angle The angle taken up by the wheel's vertical axis.
Cart spring Old-fashioned way of describing a half-elliptic leaf spring.
Coupé (or fixed-head coupé) Where the body style has a permanent metal or glass-fibre roof panel.
Drop-head coupé Body has a fold-down soft-top hood.
Fastback style When the roof line sweeps smoothly down to the extreme tail of the car.
Four-speed/five-speed transmissions The number of forward gear ratios in the gearbox.
Hatchback Opening rear body panel, which includes the rear window and provides access to luggage accommodation.
Homologation special Car produced in the minimum quantity to qualify for use in motor sport.
Live axle Means a solid beam rear axle, with the final drive and drive shafts contained inside.
Mid-engine The engine is behind the seats, but ahead of the line of the rear wheels.
Monocoque Refers to a structure acting as the chassis and also as the body shell in one welded, or moulded, whole.
Multi-tubular frame Chassis frame with many tubes to provide strength, often three-dimensional.
Overdrive An extra two-speed gearbox, mounted to the main transmission, to gear up top gear if needed.
Overhead camshaft Camshaft carried in the engine's cylinder head, operating directly on to the valve stems.
Rear engine Engine mounted in the tail, behind the line of the rear wheels.
Roadster Open two-seater sports car (see Spider).
Running gear Usually refers to the engine/gearbox/transmission of the car.

Single cam A cylinder head with only one camshaft driving the valves.
Space frame Scientifically designed multi-tube chassis frame, with all tubes in tension or compression, but not bending or in torsion.
Spider (or Spyder) Italian name for open sports car (see Roadster).
Supercar Name usually applied to the very fastest car of its period.
Torque tube Transmission shaft enclosed in a tube rigidly linking the engine to the final drive.
Transaxle Where transmission and final drive are all mounted in the same casing.
Transverse engine Line of engine crankshaft is across the car, rather than in line with its axis.
Twin-cam Refers to a cylinder head with two camshafts driving two lines of valves (the ideal arrangement).
2+2 or '+2' seating Refers to cars having small 'occasional' seats behind the front seats, not big enough for adults to use for long journeys.
Unit construction (also monocoque) Where the body/chassis structure is in one welded-up, or moulded, unit.